CLYDE W. SCOTT
P.O. BOX 507
BOTHELL, WA 98041

CLYDE W. SCOTT
P.O. BOX 507
BOTHELL, WA 98041

CLYDE W. SCOTT
P.O. BOX 507
BOTHELL, WA 98041

GENERAL ELECTRIC

ADVANCED GENERATION
DIESEL-ELECTRIC AND ELECTRIC LOCOMOTIVES

JAMES W. KERR

A
DELTA
PUBLICATION

GENERAL ELECTRIC

ADVANCED GENERATION
DIESEL-ELECTRIC AND ELECTRIC LOCOMOTIVES
— THE SECOND AND THIRD GENERATION LOCOMOTIVES —

COVER — UNION PACIFIC RAILROAD

Special thanks are extended to all those who have co-operated
and assisted to help make this publication a reality.

ISBN 0-919295-19-3

COPYRIGHT

©1989 by Delta Publications Associates Division of DPA-LTA Enterprises Inc. Printed in Canada.
All Rights Reserved. This book will not be reproduced, in whole or in part by any means, especially photocopying, including data storage and retrieval systems, without prior written permission from both the Author and Publisher. Published by Delta Publications Associates Division of DPA-LTA Enterprises Inc.

— DELTABOOKS —
AUTHOR, PRESIDENT AND CHIEF EXECUTIVE OFFICER
JAMES W. KERR

PUBLISHER
DELTA PUBLICATIONS ASSOCIATES DIVISION
DPA-LTA ENTERPRISES INC.

P.O. Box 377
Alburg, VT 05440

P.O. Box 100 - Station "R"
MONTREAL, Que., Canada
H2S 3K6

CONTENTS

	Pages
General Introduction to General Electric Current Locomotive Highlights............	4
Dash 8 Diesel-Electric Locomotive Specifications	5-17
FDL Diesel Engines and Major Components.....................................	18-33
Photographs of Popular Diesel-Electric and Electric Locomotives over the past decade using the choicest representative builders photographs available..................	34-63
COLOR SECTION (Identification Page 64)	64-80
Diesel-Electric Locomotive Specifications (C36-7, C30-7, B36-7, B30-7, B23-7 Models)......................................	81-130
Diesel-Electric Switching Locomotive Specifications (SL80, SL110, SL144 Models)	131-154
Diesel-Electric Export Locomotive Specifications (U30C, U18B Models)..........	155-173
Universal Diesel-Electric Locomotive World User List (eff. to first quarter 1987) ...	174-179
Electric Locomotive Specifications (E60CP Model)	180-199
Electric Mine Locomotive Specifications (single and double trucks)................	200-209
General Electric Locomotive Milestones	210-218
Dash 8 Engineer's Operation Manual	218-250

GENERAL INTRODUCTION TO GE CURRENT LOCOMOTIVES AND THEIR HIGHLIGHTS

General Electric Company is one of the World's largest established industrial enterprises, with annual sales greatly exceeding forty billion dollars; with more than ten thousand diesel-electric locomotives operating in 46 countries. The Leading Locomotive Builder because of commanding sales.

GE's locomotive heritage, through predecessor companies, surpasses a century of locomotive construction in which GE holds the dominant honors. Honors, like the World's first electric locomotive, invented by Thomas A. Edison in 1880. (Reference is made to our 1981 book. "Centennial Treasury of General Electric Locomotives 1880-1980". Honors, like the World's first gas-electric locomotive which quickly led to the World's first diesel-electric locomotive. Honors, like the World's first gas-turbine electric locomotive. Generally, since inception, and over the years, but prior to the late 1950's, GE was a major producer of electric locomotives, diesel-electric switcher locomotives, mine locomotives, and was a prime supplier of electrical components for ALCO diesel-electric locomotives. However, in the late 1950's, GE entered the U.S. mainline diesel-electric locomotive market, at a time which was recessionary for locomotive sales, gradually getting a foothold in a market dominated by one major builder. After years of experience, billions of locomotive miles of research and competition and about a billion dollars invested in research and testing, GE succeeded earlier in this decade in surpassing it's long established major competitor in annual sales of locomotives, and moreover, sustained this important lead.

One of the notable factors in this market penetration is a $100 Million Diesel Engine Fabrication Plant in Grove City, PA, about 50 miles South of Erie. Here, in an advanced technological environment, up to fifteen diesel engines can be almost completed weekly — the balance of completion is done at the Erie Plant.

The new Dash 8 4000hp diesel-electric locomotives can operate on high-priority express corridors, and due to their super-efficient reliability, produce power to keep mainline freight trains running on fast demanding schedules, with an outstanding record of on-time deliverance yet establishing new records in low fuel consumption.

From the Author's observation viewpoint, GE appears to have built a locomotive business foundation as solid as the Rock of Gibraltar, an enduring fortress, with dominant penetrability of the domestic diesel-electric locomotive market, due to constant listening and producing performance for the Continent's major systems. A bright sales future looms ahead if GE continues on it's present path.

James W. Kerr

CONCEPTION

From its inception in the early 1980's, our Dash 8 locomotive has been a product not just of the General Electric Company, but of the railroad industry as well. Your needs, your suggestions, and your constructive criticism shaped the design of the Dash 8. Your tests and evaluations of prototypes, demonstrators, and preproduction units refined that design. And, your operating experience with production models continues to be a major influence on every Dash 8 we build.

Today's Dash 8 is a product of that partnership, one that is designed to improve railroads' competitiveness in the transportation marketplace. The Dash 8 can help you reduce operating expense through reduced consist size while improving asset utilization through computer diagnostics, enhanced reliability, and improved maintainability. The fact is that, from the ground up, the Dash 8 is designed and built to help keep railroads as healthy on the balance sheet as they are productive out on the main line.

That's not a challenge to be met once and then forgotten. It's one that we both must work at every day. On the following pages you'll read about the locomotive itself, the manufacturing and test facilities, and the customer support that make the Dash 8 an outstanding locomotive value today. Yet to be described are the innovations that we will work together to implement in the future to keep the Dash 8 an important part of your competitive edge.

W.S. Butler
General Manager—Marketing
GE Transportation Systems

Dash 8 Performance And Availability Deliver Improved Asset Productivity

General Electric's Dash 8 locomotive delivers performance, reliability, and maintainability characteristics that can keep you competitive and strengthen your bottom line.

This new generation locomotive has been designed to reduce operating costs significantly and to provide a healthy Return on Assets.

Productivity Opportunities

▬▬▬ ▬▬▬ ▬▬▬ ▬▬▬ ▬▬▬ ▬▬▬
▬▬▬ ▬▬▬ ▬▬▬ ▬▬▬

5 DASH 8-40C LOCOMOTIVES REPLACE 6 SD40/C30-7*

▬▬▬ ▬▬▬ ▬▬▬ ▬▬▬
▬▬▬ ▬▬▬ ▬▬▬

3 DASH 8-40B LOCOMOTIVES REPLACE 4 GP40**

Unit Reductions ➡ Operating Savings

For example, the Dash 8's pulling power can deliver an immediate benefit to your railroad: the potential to replace a four-unit consist with three Dash 8's, or five units with four, can yield a significant reduction in fuel costs and other operating expenses. (See chart above).

This ability to pull higher tonnage trains, and the resulting increase in gross ton miles per unit, translates directly into improved asset productivity.

Outstanding Performance

Through Continuing Equipment Advances

Microprocessor controls work continuously to optimize Dash 8 locomotive performance while the MICROSENTRY system, which controls wheel slips, helps provide the maximum tractive effort rail conditions permit.

Microprocessor control also increases horsepower available for traction by constantly evaluating and adjusting the auxiliary system to match actual locomotive needs. This saves fuel and the additional traction horsepower results in higher speed on grades, more rapid acceleration of premium trains, and better service to your shippers.

Transitionless GE alternators provide full-time parallel traction motor operation on both four and six-axle locomotives.

Recent refinements to the GE-752™ traction motor have increased the continuous tractive effort rating of the Dash 8 locomotive substantially — up to 11% depending on the locomotive model.

*Based on test in Revenue Service on U.S. Railroad 1.4%, 7 mile grade, 28.2% adhesion.
**Estimate based on comparison of published data for horsepower, tractive effort.

A traction motor thermal protection system eliminates the need to power match and allows full use of motor capacity for faster train acceleration and reduced time on grades.

Lower Fuel Consumption

Today, average fleet fuel savings are 20% over the initial 1977 New Series locomotives — through such means as engine and turbocharger improvements and a self-contained dynamic braking package. Fuel efficiency is also enhanced by using variable speed AC motor drives for the radiator fan, equipment blowers, and the air compressor. The Dash 8 microprocessor automatically selects the minimum speed of each AC auxiliary drive to meet locomotive operating conditions.

The Dash 8 Locomotive Delivers More Power To The Rail For Each Gallon Of Fuel It Burns

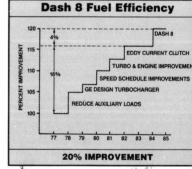

DASH 8

Outstanding Availability

From Joint Effort On Reliability

Working with the railroads, we have enhanced the over-the-road reliability of the Dash 8 to increase operator's productivity and return on assets.

As a result of this cooperative approach, refinements have been made to the GE diesel engine and support systems. For example: welded head and liner assemblies, simplified piping and fewer fittings, an improved water flow control system, and microprocessor control of engine functions. These refinements were phased in at the same time our diesel engine was accumulating more than six hundred million miles of revenue service experience at maximum power ratings.

The locomotive electrical system eliminates many mechanical relays, interlocks on power devices, and hundreds of separate wires — replacing them with more reliable microelectronic circuit cards.

VPI insulation in the Dash 8's GE-752 traction motors improves heat transfer and moisture resistance.

To deliver the mission reliability railroads asked for, the Dash 8 constantly monitors the locomotive's systems and recon-

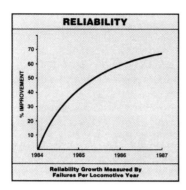

Reliability Growth Measured By Failures Per Locomotive Year

figures operation around faults if they occur. Whenever possible the locomotive continues to operate rather than shutting down, providing an extra measure of "get-home" capability to minimize road failures.

Through Faster Maintenance

On-board diagnostics and a self-test feature combine to provide a powerful troubleshooting tool to simplify and expedite maintenance by pinpointing problem areas. The Dash 8 knows what it is doing, what it is not doing, and what it should be doing; and it can report that data to maintenance personnel to speed up the troubleshooting process and help minimize maintenance or repair time.

Dash 8 equipment is arranged for fast locomotive turnaround and minimal downtime, with ready access to scheduled maintenance items and easy removal of major components.

GE's Replaceable Unit (RU) microprocessor control concept also improves maintainablility as well as availability. All microprocessor control equipment is packaged in Replaceable Units. Main excitation, auxiliary, and battery charging RUs are identical. As a result, a single replacement RU can be put into any of these three locations. The same interchangeability feature applies as well to the equipment blower and radiator fan RUs.

Through Reduced Maintenance Needs

Less time in the shop means more time on the road.

The only running maintenance required prior to 92 days is replenishment of traction motor support bearing oil and gear case lubricants, at approximately 30 day intervals depending on the locomotive's application.

Design improvements in lube oil filters, fuel filters, and other components have made the extended service intervals possible.

Dash 8 — The Locomotive Railroads Have Been Asking For

The locomotive features highlighted on these pages reflect what we've learned while working closely with the world's railroads. The end result is a locomotive that does the job you want it to do, and does it cost effectively.

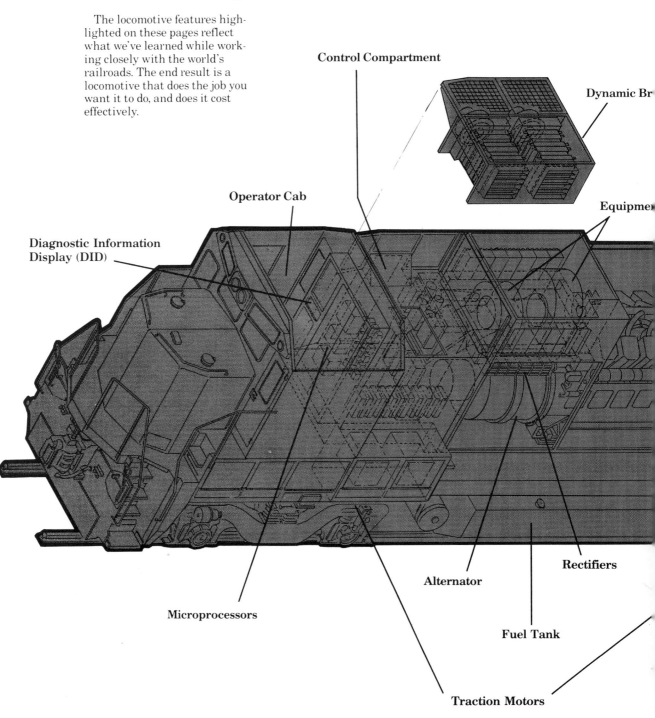

- Control Compartment
- Dynamic Br(ake)
- Operator Cab
- Equipme(nt)
- Diagnostic Information Display (DID)
- Rectifiers
- Alternator
- Fuel Tank
- Microprocessors
- Traction Motors

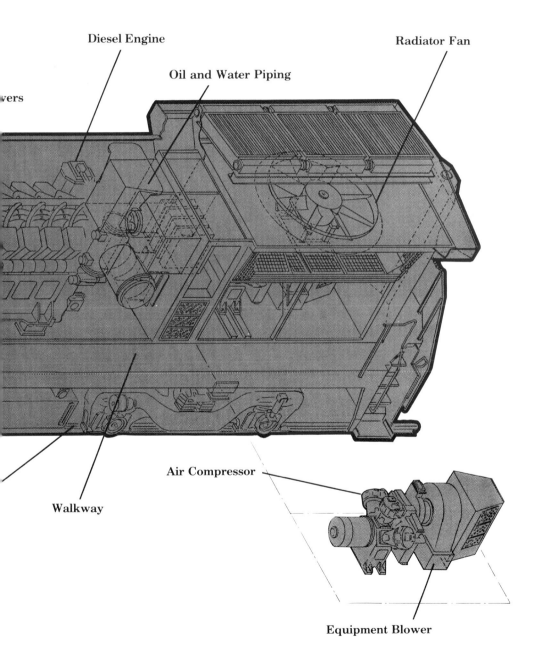

DASH 8

Control Compartment

Microprocessor and power electronics are located in one walk-in compartment for easy inspection and maintenance. Compartment is pressurized with filtered air and sealed against water leaks.

Dynamic Braking

Smaller, lighter two-stack package uses Poly Zi 300™ grids which bolt directly onto the top of the control compartment. Self-ventilated with two DC motor driven blowers. No auxiliary power needed for grid cooling.

Equipment Blowers

Computer-controlled, variable speed, AC motor-driven equipment blowers use only the auxiliary power required. Two blower arrangement provides versatility and energy savings.

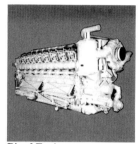

Diesel Engine

Four stroke cycle diesel engine. Improved 3-ring piston reduces lube oil consumption. Welded heads and liners, and simplified piping with fewer fittings minimize potential leaks.

Oil and Water Piping

Simplified oil and water piping, fewer fittings, and greater alignment flexibility help prevent leaks. Easy access to engine oil pump, water pump, oil cooler, water dump valve, and pipe connections.

Radiator Fan

Computer-controlled, variable speed, AC motor-driven radiator fan. Fan and motor removable through side-screen opening without removing radiators, or through top.

Air Compressor

Air-cooled, two-stage, AC motor driven air compressor eliminates engine drive shaft and couplings, as well as engine-to-compressor alignment. Easily removed from the side. Does not run when unloaded.

Walkway

Full-width walkway and recessed screens provide more room for walking or working.

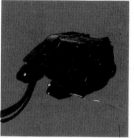

Traction Motors

Highest tractive effort rating in the industry. Thermal protection system automatically allows full utilization of motor capacity. VPI of armature and field lowers operating temperature and improves moisture resistance.

Rectifiers

Double-sided cooling of rectifiers improves heat dissipation, providing longer diode life. Each diode stack easily accessible from walkway.

Fuel Tank

Fuel tank capacity up to 5000 gallons available for longer range.

Alternator

Two alternators on a common shaft powers locomotive from zero to maximum speed with motors permanently connected in parallel, with no holes in the tractive effort curve. Auxiliary alternator eliminates DC auxiliary machines.

Microprocessors

Microprocessor control manages locomotive systems, monitors operation, and makes adjustments automatically. Works around faults to improve mission reliability. MICROSENTRY adhesion system integrated with microprocessor control.

Diagnostic Information Display (DID)

Interactive diagnostics panel in the operator's cab automatically reports faults and, on command, displays performance data requested by the operator.

Operator Cab

Seven foot ceiling, large windows, interior designed to comply with FRA noise regulations.

Dash 8 Performance And Availability Supported By Technology And Talent

Committed to making the Dash 8 locomotive the productivity standard of the industry, General Electric blends advanced technologies with talented people. Our goal is perfection in the way we build our locomotives and the way we support them in service.

Advanced Manufacturing Technology Supports Reliability Goals

On the factory floor, progress toward that goal can be measured by the hundreds of millions of dollars invested in advanced manufacturing technologies. At our "Factory with a Future" quality-oriented processes, featuring computer controls and extensive use of robotics, help assure consistent adherence to high standards of quality.

For example, the computer-controlled, completely automated Flexible Machining System for traction motor frames achieves critical dimensions and close tolerances with exceptional consistency, making a significant contribution to the reliability of the GE-752 traction motor.

Robots are linked to computer-controlled machining systems which use lasers for accurate measurements and machine feedback on crucial quality standards.

In diesel engine manufacturing an on-line operator information system on each major machine used in component production instantaneously supplies operators with work piece quality data, allowing them to make timely quality control analyses and decisions. An automated engine test facility cycles each diesel engine through a complete testing sequence, analyzing performance against design criteria and quality requirements.

One of the major features of the Dash 8 is it's concept of modular construction, allowing quality testing of components and subassemblies at progressive stages of manufacturing.

Testing of electronic assemblies, for example, uses a building-block approach to help assure product integrity. Components such as integrated circuits and power semiconductors, as well as complete microelectronic panels, are subjected to burn-in and rigorous testing. Sophisticated in-circuit testing of the panels verifies their integrity at extremes of temperature and vibration.

Computer-driven facility automatically cycles engine through complete test sequence.

The reliability and quality process culminates in the $6 million final test facility where technology, GE test operators and railroad people make a joint effort to see that you get what you are paying for. In final test, locomotives are put through their paces — in computer-monitored integrity and diagnostics testing, and in standstill dynamic testing from notch 1 through 8 to verify reliability. Some units from each order are further tested on a dedicated track to test ride, sway, speed, traction, braking, and performance.

Master computer manages Flexible Machining System.

GE test operators personally sign-off on 300 checks of the functional operation of the total locomotive system. These checks include everything from the headlight to motor rotation, from door latches to dynamic braking, from toilets to grid loading. The electrical integrity check alone includes over 1250 test steps. Throughout the testing process railroad visitors are more than welcome as partners in the final check of their locomotives.

Once the GE test operators have given their personal OK to ship, a copy of the test data sheets, including information such as tractive effort and alternator characteristic curves as verified in test, goes with the locomotive to the customer.

Greater reliability on the road is the end result of this intensive quality control at every stage of Dash 8 manufacturing.

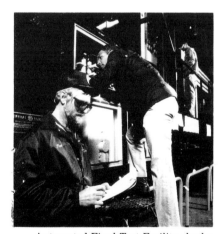

Automated Final Test Facility checks functionality and integrity of each locomotive's systems.

Dash 8 Excellence Is A Product Of People As Well As Technology

From the factory floor to the front offices, GE people are the key to applying technology to make the Dash 8 an immediate, and continuous, productive force in any fleet. GE has invested extensively to train production and test workers to make the most of the advanced manufacturing and quality control technologies used in Dash 8 production.

Beyond building for reliability, GE people tap powerful computer networks and data communications links to help keep Dash 8 locomotives productive.

GE replacement parts experts already have a wealth of experience serving railroad customers. Now they're aided by a new information system that helps them provide immediate information on material availability, price, and delivery. Same-day order entry is a reality, and parts can often be shipped within 24 hours. GE replacement parts specialists can also use the system to help railroads plan for their parts requirements to control inventory as well as to design component overhaul kits to meet specific needs.

Application and product service engineers use computer technology with advanced communications capability to keep Dash 8 fleets in service. Whether to perform application studies to determine the Dash 8 configuration that best meets your specific needs, to put it in service, or to keep it on the road, our technical support teams have access to product and performance data bases that supplement their own experience and training. GE field service engineers working with you can draw on that data base to help improve performance and manage maintenance.

Add to that the Dash 8's own on-board computer, with it's diagnostic and performance record keeping capabilities, and you have a package of people and technology that works hard to achieve maximum productivity through superior locomotive availability and performance. There are yet more people at GE dedicated to helping you make the most of the Dash 8. There are trainers to teach effective maintenance practices, experts who can help your people learn to use the Dash 8 computer system to maximum effect, and technical writers and catalog specialists who provide additional information to go with your Dash 8's.

Operational data is transferred from locomotive computer memory to serve engineer's records.

The fact is, Dash 8 performance and availability are the product of dedicated, talented people supported by advanced manufacturing and computer technology to build a locomotive that works hard for you.

Performance Characteristics

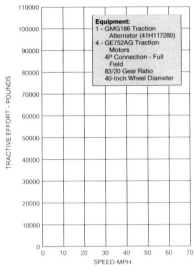

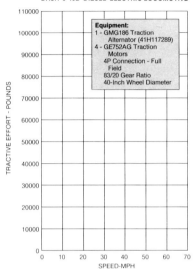

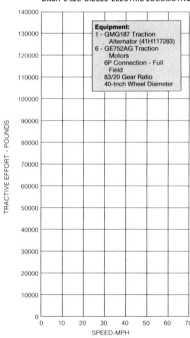

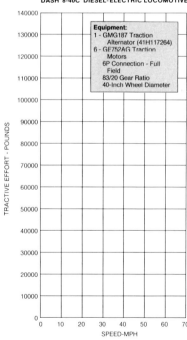

General Electric Dash 8 Locomotives General Characteristics

	DASH 8 - 32B	DASH 8 - 32C	DASH 8 - 40B	DASH 8 - 40C
GE Model				
AAR Class	General Purpose	General Purpose	General Purpose	General Purpose
Wheel Arrangement	B - B	C - C	B - B	C - C
Major Dimensions				
Length	63 ft. 7 in.	67 ft. 11 in.	66 ft. 4 in.	70 ft. 8 in.
Height	14 ft. 11½ in.	15 ft. 4½ in.	14 ft. 11½ in.	15 ft. 4½ in.
Width	10 ft. 2¾ in.	10 ft. 2¾ in.	10 ft. 2¾ in.	10 ft. 2¾ in.
Bolster Centers	36 ft. 7 in.	40 ft. 7 in.	40 ft. 1½ in.	43 ft. 4 in.
Truck Wheel Base	9 ft. 0 in.	13 ft. 7 in.	9 ft. 0 in.	13 ft. 7 in.
Wheel Diameter	40 in.	40 in.	40 in.	40 in.
Minimum Curvature (ft./degree)				
Single	150'/39°	273'/21°	150'/39°	273'/21°
Multiple Unit	195'/29°	273'/21°	195'/29°	273'/21°
Weight (lbs.)				
*Maximum (nominal ±2% tolerance)	284,000	390,000	288,000	410,000
On Drivers	284,000	390,000	288,000	410,000
Gear Ratio				
Standard - Maximum Speed	83/20-70	83/20-70	83/20-70	83/20-70
Optional - Maximum Speed	81/22-75	—	81/22-75	—
Maximum Continuous Tractive Effort and Speed				
Standard (lbs./mph)	71,600/13.8	109,700/8.2	69,200/18.6	108,600/11.0
Optional	63,500/15.6	—	61,400/20.9	—
Supplies				
Fuel (gals.)	3250	3900	3250	5000
Water (gals.)	350	350	380	380
Lube Oil (gals.)	365	365	410	410
Sand (cu. ft.)	48	48	48	48
Operating Station	Single	Single	Single	Single
Control	Microprocessor	Microprocessor	Microprocessor	Microprocessor
Draft Gear	NC390	NC390	NC390	NC390
Engine Data				
Model	FDL-12	FDL-12	FDL-16	FDL-16
Engine Type	V-12 - 4 Cycle Turbocharged	V-12 - 4 Cycle Turbocharged	V-16 - 4 Cycle Turbocharged	V-16 - 4 Cycle Turbocharged
Cylinders	12	12	16	16
Traction Horsepower	3200	3200	4000	4000
Bore and Stroke	9 x 10½ in.	9 x 10½ in.	9 x 10½ in.	9 x 10½ in.
RPM (maximum)	1050	1050	1050	1050
Compression Ratio	12.7:1	12.7:1	12.7:1	12.7:1
Traction Equipment				
Traction Alternator and Auxiliary Alternator	GMG 186	GMG 187	GMG 186	GMG 187
Traction Motors	4 752AG	6 752AG	4 752AG	6 752AG
Auxiliary Equipment				
Radiator Fan Motor (AC)	1 GYA30A	1 GYA30A	1 GYA30A	1 GYA30A
Traction Motor Blowers (AC)	1 GDY75C #1 End	1 GDY76C #1 End	1 GDY75C #1 End	1 GDY76C #1 End
	1 GDY75E #2 End	1 GDY76E #2 End	1 GDY75E #2 End	1 GDY76E #2 End
Alternator Blower	1 GDY74B	1 GDY74D	1 GDY74B	1 GDY74D
Air Brake Schedule	26L	26L	26L	26L
Air Compressor				
Motor (AC)	1 GYA28A	1 GYA28A	1 GYA28A	1 GYA28A
Drive/Control	AC - Microprocessor	AC - Microprocessor	AC - Microprocessor	AC - Microprocessor
Capacity	236 CFM	236 CFM	236 CFM	236 CFM
Equipment Air Filtration				
Primary	Vortex	Vortex	Vortex	Vortex
Secondary	Fiberglass	Fiberglass	Fiberglass	Fiberglass

*With heavy options and/or maximum ballast.

DASH 8

Location of Apparatus

1. Sand Fill
2. Toilet Area
3. Sand Box
4. Handbrake
5. Refrigerator or Cooler
6. Emergency Brake Valve
7. Heater and Defroster
8. Headlight and Number Light Box
9. Control Area #5 (Control Console)
10. Heater, Side Strip
11. Engine Control Panel
12. Trucks:
    "B" - 2 Axles Per Truck
    "C" - 3 Axles Per Truck
13. Control Area #1
14. Control Area #2 ⎫ Located In
15. Control Area #3 ⎬ Auxiliary
16. Control Area #4 ⎭ Cab
17. Control Area #6
18. Air Brake Compartment
19. Control Area #7
20. Control Area #8
21. Battery Box
22. Dynamic Braking Box
23. Rectifiers (Propulsion) and Fuses
24. Blower Box and Air Filters
25. Alternators (Main and Auxiliary)
26. Engine Start Station
27. Engines:
    32B/32C - 12 Cyl. FDL-12
    40B/40C - 16 Cyl. FDL-16
28. Fuel and Retention Tanks
29. Fuel Fill
30. Fuel Gage
31. Lube Oil Cooler
32. Lube Oil Filter
33. Engine Water Tank and Water Control Valve
34. Engine Air Filter Compartment
35. Air Compressor (Motor Driven)
36. Control Area #9
37. Radiators
38. Radiator Fan
39. Blower and Air Filters (No. 2 End)
40. Anti-Climber (Optional)
41. Pilot Plate (Snow Plow Optional)

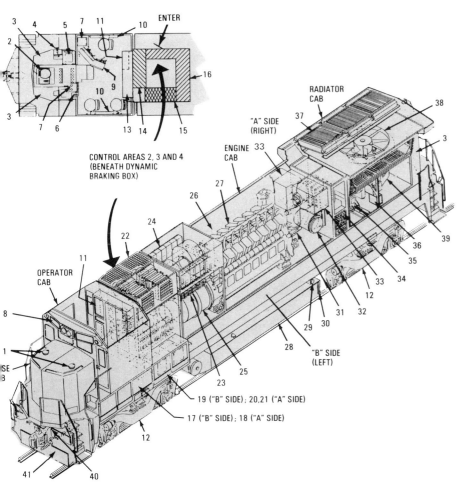

-15-

GENERAL ELECTRIC DASH 8 LOCOMOTIVES

General Characteristics

	B23-8	B32-8	C32-8	B39-8	C39-8
ENGINE					
Type	GE FDL-12	GE FDL-12	GE FDL-12	GE FDL-16	GE FDL-16
Horsepower - Traction	2,300	3,170	3,150	3,900	3,900
Number of Cylinders	12	12	12	16	16
Bore and Stroke	9 x 10½ in	9 x 10½ in	9 x 10½ in	9 x 10½ in	9 x 10½ in
R.P.M.	1,050	1,050	1,050	1,050	1,050
Compression Ratio	12.7:1	12.7:1	12.7:1	12.7:1	12.7:1
Cycle	4	4	4	4	4
Turbocharged	Yes	Yes	Yes	Yes	Yes
Engine Cooling Fan and AC Motor	1	2	2	2	2
Static Engine Cooling Fan Drive	Variable Speed	Variable Speed	Variable Speed	Variable Speed	Variable Speed
OPERATING CAB & CONTROLS	General purpose	General purpose	General purpose	General purpose	General purpose
WHEEL ARRANGEMENT	B-B	B-B	C-C	B-B	C-C
JOURNAL BEARINGS					
Type	Grease/tapered roller	Grease/tapered roller	Grease/tapered roller	Grease/tapered roller	Grease/tapered roller
Class and size	GG (6⅞)	GG (6⅞)	GG (6⅞)	GG (6⅞)	GG (6⅞)
TRACTION EQUIPMENT					
Main Alternator	GMG 186	GMG 186	GMG 187	GMG 186	GMG 187
Traction Motor	4-GE752	4-GE752	6-GE752	4-GE752	6-GE752
Traction Motor Blowers and AC Motors	2	2	2	2	2
Static Blower Drive	Variable Speed	Variable Speed	Variable Speed	Variable Speed	Variable Speed
Wheelslip Correction	GE MICROSENTRY	GE MICROSENTRY	GE MICROSENTRY	GE MICROSENTRY	GE MICROSENTRY
AIR BRAKE SCHEDULE	26L	26L	26L	26L	26L
MAJOR DIMENSIONS					
Length	63 ft 7 in	63 ft 7 in	67 ft 11 in	66 ft 4 in	70 ft 8 in
Height	14 ft 11½ in	14 ft 11½ in	15 ft 4½ in	14 ft 11½ in	15 ft 4½ in
Width	10 ft 1¾ in	10 ft 1¾ in	10 ft 1¾ in	10 ft 1¾ in	10 ft 1¾ in
Bolster Centers	36 ft 7 in	36 ft 7 in	40 ft 7 in	39 ft 4 in	43 ft 4 in
Truck Wheel Base	9 ft 0 in	9 ft 0 in	13 ft 7 in	9 ft 0 in	13 ft 7 in
Minimum Track Curvature Rad. & Deg.					
(1) For Single Unit	150'/39°	150'/39°	273'/21°	150'/39°	273'/21°
(2) For MU	195'/29°	195'/29°	273'/21°	195'/29°	273'/21°
DRIVING WHEEL DIAMETER	40"	40"	40"	40"	40"
WEIGHT*					
On Drivers - # Min. & Max.	254,300/280,000	262,900/280,000	346,300/420,000	274,000/280,000	365,100/420,000
Total Minimum and Maximum	254,300/280,000	262,900/280,000	346,300/420,000	274,000/280,000	365,100/420,000
TRACTIVE EFFORT					
Starting at 25% adhesion for Minimum and Maximum Weight	63,575/70,000	65,725/70,000	86,575/105,000	68,500/70,000	91,275/105,000
Continuous Tractive Effort and Speed (mph)					
(1) For 70 mph gearing	71,890/9.2	70,140/13.9	108,360/8.2	68,100/18.3	106,790/10.9
(2) For 75 mph gearing	63,780/10.4	62,230/15.7	96,140/9.2	60,420/20.6	94,740/12.3
GEAR RATIO AND MAXIMUM SPEED (mph)					
Standard	83/20-70	83/20-70	83/20-70	83/20-70	83/20-70
Optional	81/22-75	81/22-75	81/22-75	81/22-75	81/22-75
SUPPLIES					
Fuel for Min. & Max. Tank, Gallon	2,150	3,150	3,900	3,150	3,900
Coolant, Gallon	350	350	350	380	380
Lube Oil, Gallon	260	260	260	360	360
Sand, Cubic Feet	40	40	40	40	40
COMPRESSOR, AIR CFM (MAXIMUM)	296	296	296	296	296
DRAFT GEAR	NC 390	NC 390	NC 390	NC 390	NC 390
AIR FILTERING DEVICES					
Primary	Vortex Self Clean	Vortex Self Clean	Vortex Self Clean	Vortex Self Clean	Vortex Self Clean
Secondary Engine Air Intake	GE Paper	GE Paper	GE Paper	GE Paper	GE Paper
Engine Room Pressurized	Yes	Yes	Yes	Yes	Yes
Main Generator Pressurized	Yes	Yes	Yes	Yes	Yes

*Weight subject to manufacturing tolerance of ±2%. Modifications may increase weight.

Location of Apparatus (B39-8)

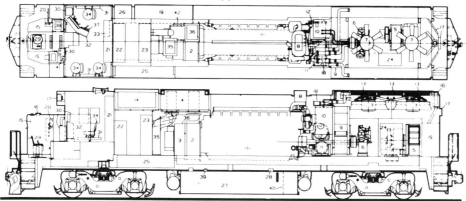

LEGEND

1. Engine GE Model 7FDL16
2. Alternator
3. Auxiliary Alternator
4. No. 1 End Blower Package
5. No. 2 End Blower Package
6. Air Compressor
7. Radiator Fan
8. Engine Muffler
9. Engine Air Filter Compartment
10. Engine Water Tank
11. Lube Oil Cooler
12. Lube Oil Filter
13. Radiator
14. Dynamic Braking Box
15. Sand Box
16. Sand Filler
17. Headlight/No. Light Box
18. Fluid Amplifier
19. Battery Box
20. Hand Brake
21. Control Compartment No. 1
22. Control Compartment No. 2
23. Control Compartment No. 3
24. Control Compartment No. 4
25. Control Compartment No. 7
26. Control Compartment No. 8
27. Fuel Tank
28. Fuel Filler
29. Toilet
30. Electric Heater & Defroster
31. Side Strip Heater
32. Control Console
33. Engine Control Panel
34. Sliding Seat
35. Air Duct - Traction Motor Blower
36. Air Duct - Alternator
37. Air Brake Valve
38. Refrigerator
39. Fuel Gage
40. Retension Tank
41. Emergency Brake Valve
42. Engine Start Station

Performance Characteristics

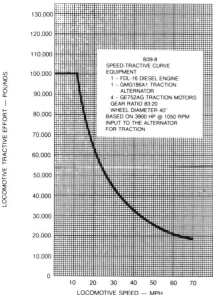

B39-8
SPEED-TRACTIVE CURVE
EQUIPMENT
1 – FDL-16 DIESEL ENGINE
1 – GMG186A1 TRACTION ALTERNATOR
4 – GE752AG TRACTION MOTORS
GEAR RATIO 83:20
WHEEL DIAMETER 40"
BASED ON 3900 HP @ 1050 RPM INPUT TO THE ALTERNATOR FOR TRACTION

Performance Characteristics

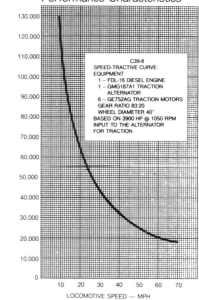

C39-8
SPEED-TRACTIVE CURVE
EQUIPMENT
1 – FDL-16 DIESEL ENGINE
1 – GMG187A1 TRACTION ALTERNATOR
6 – GE752AG TRACTION MOTORS
GEAR RATIO 83:20
WHEEL DIAMETER 40"
BASED ON 3900 HP @ 1050 RPM INPUT TO THE ALTERNATOR FOR TRACTION

Reprinted from the 1984 Car and Locomotive Cyclopedia Copyright: Simmons-Boardman Publishing Corp.

FDL SERIES

General Electric's diesel engine reliably provides high horsepower at low operating cost.

General Electric Diesel Engine
...reliable service with low fuel and maintenance costs

General Electric's advanced four-cycle diesel engine, available in 8, 12, or 16-cylinder configurations, provides rugged power and minimum operating costs for the toughest applications. This engine combines high technology with practical experience.

The high-capacity turbocharger and efficient combustion system help reduce fuel consumption.

Easy accessibility to components and maximum standardization of parts reduce maintenance and spare parts inventory.

Technical assistance and service support are available from General Electric's staff of experienced application engineering and service personnel. Installation supervision, thorough operating and maintenance training, conveniently located parts centers, and a practical unit exchange program all contribute to the profitable performance of the GE diesel engine.

Unitized Cylinder
...is easily removed for inspection, repair, or maintenance

Unitized cylinders are a unique feature of the General Electric diesel engine. Each is held on top of the main frame with four bolts, and has individual water and air passages, completely isolated from the main frame.

Three major elements — the liner, steel head insert, and external jacket — make up the unitized cylinder.

The cylinder can be removed for inspection, maintenance, and repair without disturbing the piston and related wrist-pin, the connecting rod bearings, or the other cylinders.

Inspection, maintenance, and replacement of pistons is also simplified. With the cylinder removed, the piston is completely exposed above the engine frame. Piston or ring work can be done without disturbing the crankpin bearing.

Unitized cylinders on all GE diesel engines are interchangeable.

Liner is perfectly round. No ribs or projections complicate handling or accumulate heat-building deposits.

External jacket is designed so that the liner and steel head insert can be easily removed for bench maintenance.

FDL SERIES

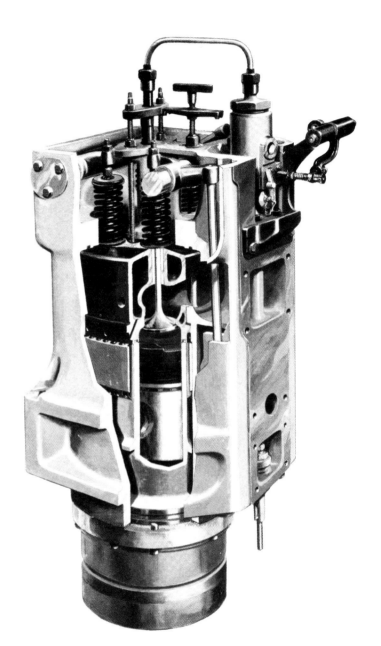

Steel head insert provides high strength and cooler operating temperatures with lower thermal stresses. Each valve has its own rotator.

Unitized cylinder bolts to the main frame. Sections of the air intake manifold, exhaust manifold, inlet and discharge water headers, and fuel supply piping connect directly to each assembly.

Precision-Engineered Engine

...features the industry's most modern design concepts

A highly-reliable exhaust manifold, constructed of stainless steel, provides unrestricted gas flow and cooler operating temperatures for long life of components in the combustion stream.

The low-pressure fuel system, delivering fuel to the two banks of cylinders in parallel, achieves high reliability and maintainability. Large hoses, together with the parallel connection, prevent starving any cylinders for fuel under full load. Banjo fittings at each pump inlet connection resist vibration and leaks.

Large 18-mm double helix pumps inject fuel within a minimum period of time and vary the timing at different throttle notches to be optimum.

Two camshafts of forged high-carbon steel with hardened cam and journal surfaces operate the engine valves and fuel injection pumps. Fully sectionalized for easy change-out, they feature identical sections (one per cylinder) on each shaft assembly.

Bearing shells are of the steel backed tri-metal type. Both main and connecting rod bearings are grooveless, resulting in a thick oil film, low load per unit area, and high lube oil header pressure. In addition, this design is tolerant of microscopic dirt particles and degraded oil conditions.

One piece silchrome 15-degree intake valves reduce scrubbing action between the valve and the seat, minimizing wear and lengthening service life. Exhaust valves are two-piece with high temperature material in the heads.

The steel crown piston provides high strength and heat resistance in the crown area and has a light weight aluminum skirt. Only two compression rings are used, resulting in long life for both piston and rings. The crown is slightly reduced in diameter above the upper ring to minimize carbon build-up and lube oil consumption.

A layshaft lever permits manual increase of fuel while cranking thus assuring quick engine starts. The engine is protected from overspeed during start-up because the overspeed link is located between the layshaft lever and the fuel pump racks.

Intake valves have 15° seats to minimize scrubbing action.

Steel crown piston provides heat and wear resistance.

Overspeed system design results in quick engine starting.

FDL SERIES

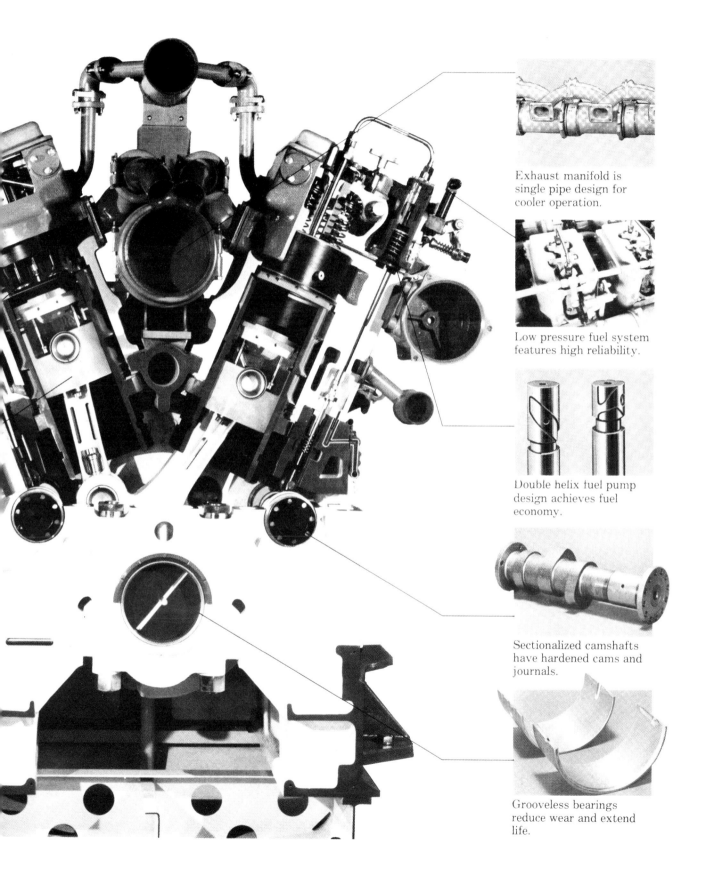

Exhaust manifold is single pipe design for cooler operation.

Low pressure fuel system features high reliability.

Double helix fuel pump design achieves fuel economy.

Sectionalized camshafts have hardened cams and journals.

Grooveless bearings reduce wear and extend life.

FDL SERIES

Dimensions And Specifications

...FDL-8 Specifications	Full Load Speed	1050 rpm	*Rotation	Counterclockwise
			Firing Order	1R-1L-2R-2L-4R-4L 3R-3L
	Idle Speed	450 rpm	Aspiration	Turbo Supercharged
	Number of Cylinders	8		
	Cylinder Arrangement	45°V	**Dimensions**	
	Stroke Cycle	4		
	Bore	9 inches	**Height	7 feet, 2¼ inches
	Stroke	10½ inches	Length	10 feet, 8½ inches
	Compression Ratio	12.7 to 1	Width	5 feet, 8¼ inches
			Weight (dry)	27,000 pounds
...FDL-12 Specifications	Full Load Speed	1050 rpm	*Rotation	Counterclockwise
			Firing Order	1R-1L-5R-5L-3R-3L- 6R-6L-2R-2L-4R-4L
	Idle Speed	450 rpm	Aspiration	Turbo Supercharged
	Number of Cylinders	12		
	Cylinder Arrangement	45°V	**Dimensions**	
	Stroke Cycle	4		
	Bore	9 inches	**Height	7 feet, 6⅛ inches
	Stroke	10½ inches	Length	13 feet, 3½ inches
	Compression Ratio	12.7 to 1	Width	5 feet, 8⅜ inches
			Weight (dry)	35,000 pounds
...FDL-16 Specifications	Full Load Speed	1050 rpm	*Rotation	Counterclockwise
			Firing Order	1R-1L-3R-3L-7R-7L- 4R-4L-8R-8L-6R-6L- 2R-2L-5R-5L
	Idle Speed	450 rpm	Aspiration	Turbo Supercharged
	Number of Cylinders	16		
	Cylinder Arrangement	45°V	**Dimensions**	
	Stroke Cycle	4		
	Bore	9 inches	**Height	7 feet, 6⅛ inches
	Stroke	10½ inches	Length	16 feet, 1 inch
	Compression Ratio	12.7 to 1	Width	5 feet, 8⅜ inches
			Weight (dry)	43,500 pounds

*Viewed from generator end of engine
**Measured from bottom of oil pan to top of intercoolers. Does not include stack.

FDL SERIES

Rugged Cast Main Frame
...incorporates integral lube passages

The main frame is designed to be simply a support for the crankshaft, camshafts, and cylinders. No water, combustion air, or exhaust gas comes in contact with it, thus minimizing corrosion and leakage. Cast iron was chosen as material for this frame for two reasons: First, it provides exceptional dimensional stability; second, it has excellent vibration dampening characteristics.

Integral lube passages in the frame eliminate separate oil lines with their susceptibility to vibration damage, loosening, or being forgotten at overhaul.

The crankcase is easily accessible through combination quick-removal and pressure-relief doors along both sides of the engine.

FDL SERIES

A Commitment Of Resources And Technology

More than $500 million has been invested in manufacturing and support facilities. State-of-the-art, highly automated machine tools have been installed in the Grove City engine manufacturing facility, where floor space has been quadrupled. The new 40,000 square foot Learning and Communications Center is dedicated to technological training of both customers and GE employees. And the Dash 8 locomotive with its microcomputer controls is setting new standards of locomotive productivity. Major contributions in reliability, fuel economy, horsepower, and maintenance simplicity come from the FDL engine.

The General Electric Commitment To Quality And Service

Components are thoroughly inspected and approved prior to installation on the engine.

The completed GE diesel engine undergoes exhaustive testing. Each is completely run-in under full load at the factory to locate and eliminate potential problems and is ready for immediate full power service.

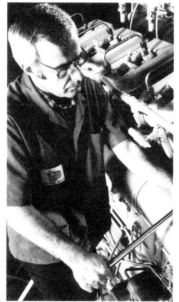

GE service programs are tailored to meet your maintenance needs. Service engineers are readily available.

Parts and service for General Electric diesel engines are available 24 hours a day, seven days a week.

Advanced design concepts originate from GE aerospace technology.

General Electric high-performance turbochargers make a significant contribution to the recognized fuel economy of GE diesel-engines. Totally exhaust driven and specifically designed for engine compatibility, they combine maximum operating efficiency with practical maintenance advantages.

Designed for railroad service, the turbos have been proven by both exhaustive laboratory testing and years of field operation. These rugged units also provide enhanced efficiency on stationary power and marine applications.

- Maximum horsepower—The 3940 gross horsepower of General Electric's advanced 16-cylinder turbocharged diesel-engine equals the potential output of an otherwise comparable 37-cylinder engine without turbocharging.

- Minimum fuel consumption—It is commonly accepted that turbocharged four-cycle diesel-engines possess an inherent fuel economy advantage over competitive engines. This advantage is increased by state-of-the-art advances in General Electric's nozzle ring, turbine disc, compressor wheel and diffuser designs.

- Operating dependability—General Electric turbochargers are exhaust driven at all speeds. They receive cooling water and lubricating oil directly from the engine systems. Improved component designs, plus elimination of separate pumps and filters enhance operating reliability. Lower pre-turbine temperatures, made possible by the design, also contribute to turbocharger durability and life expectancy.

- Maintenance savings—Design simplicity and engine compatibility improve maintenance accessibility. Field reports indicate that GE turbocharger changeouts can be accomplished much faster than gear driven turbochargers.

Because of the turbocharger's inherent reliability, the necessity of changeout is greatly reduced.

What they do

GE turbochargers are centrifugal air compressors. Driven by engine exhaust gases, they produce compressed air for two important functions:

- During the latter part of the exhaust stroke and early part of the intake stroke, there is an interval in the cycle, called the valve overlap, when both inlet and exhaust valves are open. During the overlap period, compressed air from the turbocharger is directed into the combustion chamber. This

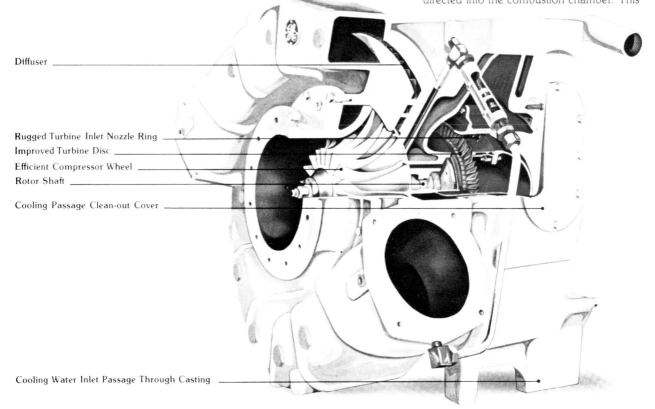

Diffuser

Rugged Turbine Inlet Nozzle Ring
Improved Turbine Disc
Efficient Compressor Wheel
Rotor Shaft

Cooling Passage Clean-out Cover

Cooling Water Inlet Passage Through Casting

compressed air expels the spent exhaust gases while lowering the temperature of piston and cylinder parts.

- Compressed combustion air contains more oxygen per unit volume than naturally aspirated combustion air. Based upon a consistent fuel-to-air ratio, the denser compressed air allows efficient burning of more fuel per power stroke.

How they do it so efficiently

Engine exhaust gases enter the turbocharger through the turbine inlet assembly. Directed by the nozzle ring blades, the hot gases impinge upon and expand through the turbine bucket blades to drive the rotor shaft. Engine air drawn from outside the engine is compressed and discharged by the compressor wheel mounted on the output end of the rotor shaft. A diffuser converts air velocity into pressure enroute to the blower casing outlets. Air density is further increased by two externally mounted air-to-water intercoolers located ahead of the engine air intake manifold.

Turbocharger lubrication is provided directly from the engine main header through a pressure-reducing orifice.

Cooling water from the engine water pump enters the turbocharger through holes in the turbocharger mounting feet and circulates throughout strategically-located cored passages in the casing.

A seal air system delivers air from the compressor to the outboard side of the oil seals. This prevents oil from bypassing the seal and reduces carbon build-up. The same seal air also provides compressed air to cool the turbine disc and creates an equalizing force which minimizes the exhaust-gas-produced load on the turbine-end thrust bearing.

Where they can be applied

A variety of 14 and 16-inch General Electric turbochargers are available for retrofit on virtually all existing 8, 12, and 16-cylinder engines. Selected existing turbocharger components and assemblies can be replaced with new turbocharger elements. Specifically, these include the new turbine inlet assembly with O.D. and I.D. clamped nozzle ring, the precision turbine disc assembly, and the high-performance rotor shaft thrust bearings.

Changeover on existing equipment requires no major engine modifications. The exhaust manifold connection and intercoolers remain unchanged. Direct engine mounting with engine-aligned internal cooling water passages eliminate turbocharger mounting brackets with extended piping.

Added maintenance advantages include minimum components, easier access, and large cooling passage cleanout covers rather than knuckle-busting plugged access holes.

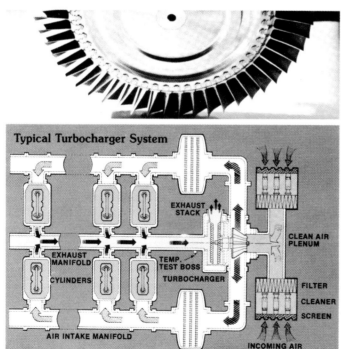

The Spin Clean System
A cost-effective modular solution to high volume primary air cleaning requirements

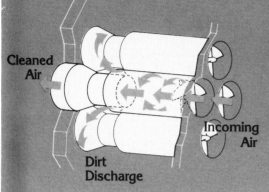

The basic spin clean module is a compact high-density molded polypropylene building block containing 54 tubes with stationary air spinners.

These scientifically-designed spinners impart high radial velocity to the dirty input air. The resulting centrifugal force hurls dirt particles and water to the periphery of the tube.

Approximately 90 percent of the inlet air passes easily through the clean air core at the center of the tube. Water on the surface of the tube and dirt concentrated in the 10 percent periphery air is withdrawn and exhausted through the bleed slots.

Dirt particles impinging on the walls of the cleaner provide an automatic self-cleaning action which prevents dirt from accumulating in the bleed slots and thus guards against plugged discharge openings.

In order to keep the spin cleaner distortion free and dimensionally stable, General Electric developed a special manufacturing facility, utilizing low-pressure transfer-injection technology. This eliminates molded-in stresses common to other injection molding processes.

Special molding process eliminates distortion from thermal cycling, thus assuring that design efficiencies are maintained throughout the long service life of the cleaner.

Diverse mounting arrangements provide application versatility

Systems capability of the spin clean module can meet a variety of application requirements. The primary spin cleaner module can be mounted in any of the three positions shown.

Three mounting positions increase application flexibility.

Multiple cleaner modules can be mounted in parallel to meet air flow requirements of the specific application. Secondary disposable filters can be used in conjunction with the spin clean system where extra fine particle removal is necessary.

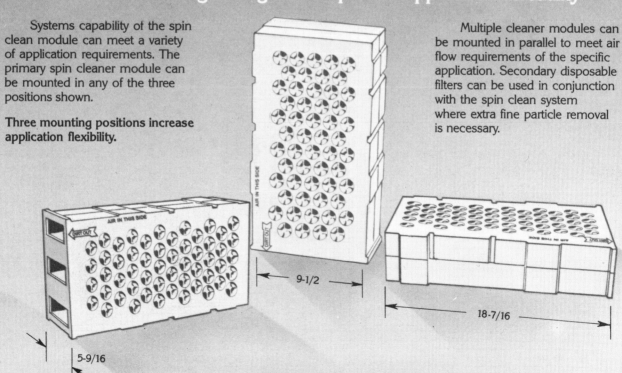

Years of service and lab testing confirm superiority

Many years of locomotive air filtration service prove spin clean performance.

Spin clean systems have provided many years of reliable, cost-effective service as the primary engine air filters on General Electric locomotives.

Through extensive laboratory testing, General Electric has evaluated performance of the spin cleaner with competitive cleaners constructed of steel, aluminum, and nylon-mellamine. The spin clean module has proven superior to these competitive cleaners.

Resistance to weight loss from abrasion is excellent. Abrasive particles actually bounce off the surface rather than scratch the cleaner. The cleaner effectively resists corrosion and deterioration from chemical agents.

The molded polypropylene construction is rugged and not subject to damage in shipment and handling.

A series of important performance criteria

The spin clean module is capable of cleaning 1200 to 2000 cfm of air at a pressure drop of two to five inches of water. It has a dirt removal efficiency of 98% for particles 15 microns and larger. The cleaner will remove more than 93% of all particles eight microns and larger. Figure 1 compares dirt removal efficiency of the cleaner with particle size.

Water removal efficiency of the spin clean module is more than 90%. With the spin design, water is forced to the outer periphery of the tubes and discharged with the bleed air.

Figure 2 illustrates the efficiency of the spin clean module through a wide range of inlet air flows. These results are based on AC standard coarse test dust with particle sizes ranging from sub-micron to 200 microns. Figure 3 shows pressure drop of 1.1 inches of water at an air flow of 1000 cfm.

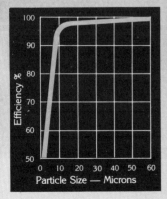

Figure 1 — Spin clean dirt removal efficiency at various particle sizes.

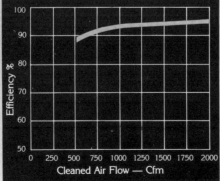

Figure 2 — Spin clean efficiency at various inlet air flows. (AC Standard Coarse Test Dust with particle sizes from sub-micron to 200 microns.)

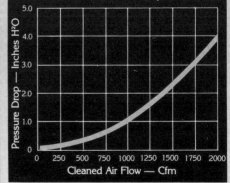

Figure 3 — Pressure drop of 1.1 inches of water at 1,000 cfm air flow.

Atchison, Topeka & Santa Fe #7401 Model B39-8 3900hp 1985

Atchison, Topeka & Santa Fe #7401 Model B39-8 3900hp 1985

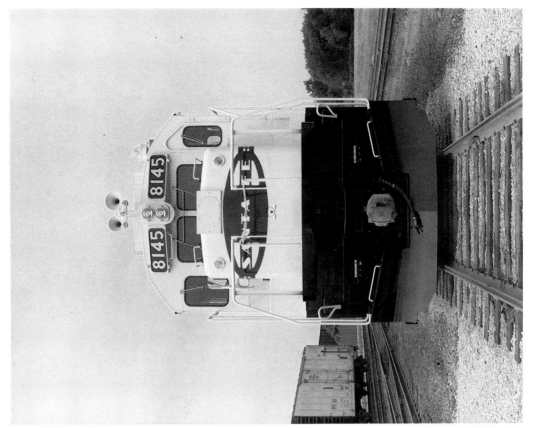

Atchison, Topeka & Santa Fe #8145 Model C30-7 3000hp 1981

Atchison, Topeka & Santa Fe #8145 Model C30-7 3000hp 1981

Burlington Northern #5497 Model B32-8 3170hp 1984

Burlington Northern #5497 Model B32-8 3170hp 1984

Burlington Northern #5497 Model C32-8 3150hp 1984

Burlington Northern #5498 Model C32-8 3150hp 1984

Burlington Northern #4000 Model B30-7A 3000hp Cabless Booster Unit 1982

Burlington Northern #4059 Model B30-7A 3000hp Cabless Booster Unit 1982

Burlington Northern #5071 Model B30-7A (12 Cyl) 3000hp 1980

Burlington Northern #5071 Model B30-7A (12 Cyl) 3000hp 1980

Burlington Northern #5071 Model B30-7A (12 Cyl) 3000hp 1980

Burlington Northern #5071 Model B30-7A (12 Cyl) 3000hp 1980

Burlington Northern #5112 Model C30-7 3000hp 1980

Chessie System (CSX) #8238 Model B30-7 3000hp 1981

Conrail #5048 Model B36-7 3600hp 1983

Conrail #5048 Model B36-7 3600hp 1983

Conrail #6588 Model C30-7A (12 Cyl) 3000hp 1984

Conrail #6588 Model C30-7A (12 Cyl) 3000hp 1984

Conrail #6610 Model C32-8 3150hp 1984

Conrail #6610 Model C32-8 3150hp 1984

CSX (Seaboard System) #5856 Model B36-7 3600hp 1985

CSX (Seaboard System) #5856 Model B36-7 3600hp 1985

CSX (Seaboard System) #3217 Model MATE (Motors for Additional Tractive Effort)

General Electric Demonstrator #505 Model C36-7 3600hp

General Electric Demonstrator #808 Model B39-8 3900hp 1986

General Electric Demonstrator #809 Model B39-8 3900hp 1988

LMX #8510 Model B39-8 3900hp 1987

-48- LMX #8517 Model B39-8 3900hp 1987

Mexico — Ferrocarril Chihuahua al Pacifico #435 Model C30-7 3000hp 1981

Mexico — Ferrocarril Chihuahua al Pacifico #435 Model C30-7 3000hp 1981

Mexico — Nacional De Mexico #9180 Model B23-7 2250hp 1981

Mexico — Nacional De Mexico #9180 Model B23-7 2250hp 1981

Norfolk Southern #8560 Model C39-8 3900hp 1984

Norfolk Southern #8560 Model C39-8 3900hp 1984

Norfolk Southern #8612 Model C39-8 3900hp 1985

Norfolk Southern #8612 Model C39-8 3900hp 1985

Norfolk Southern #8560 Model C39-8 3900hp 1987

Norfolk Southern #8688 Model C39-8 3900hp 1987

Norfolk Southern #8688 Model C39-8 3900hp 1987

Norfolk Southern #8688 Model C39-8 3900hp 1987

Southern (NS) #3820 Model B36-7 3600hp 1981

Southern Pacific #8036 Model Dash 8-40B 4000hp 1987

Southern Pacific #7769 Model B36-7 3600hp 1984

Southern Pacific #7769 Model B36-7 3600hp 1984

Union Pacific #9100 Model Dash 8- 40C 4000hp 1987 George R. Cockle Photo.

Union Pacific #9199 Model Dash 8-40C 4000hp 1988 George R. Cockle Photo.

General Electric Demonstrator Model SL110 110-Ton Switching Locomotive

Central Hudson Power #Donald F. Dupay Model SL110 110-Ton Switching Locomotive

People's Republic of China #0014 Model C36-7 rated 4000hp 1984

People's Republic of China #0022 Model C36-7 rated 4000hp 1984

Deseret-Western #WFU2 Model E60C 6000hp Electric Freight.

Mexico — Nacional De Mexico #EA001 Model E60C 6000hp Electric Pass/Frt.

Deseret-Western #WFU2 Model E60C 6000hp Electric Freight Locomotive

Mexico — Nacional De Mexico #EA001 Model E60C 6000hp Electric Pass/Frt.

National Railroad Passenger Corp (AMTRAK) #974 Model E60CP 6000hp Elect. Pass

Texas Utilities #2306 Model E25B 2500hp Electric Remote-Control Loco.

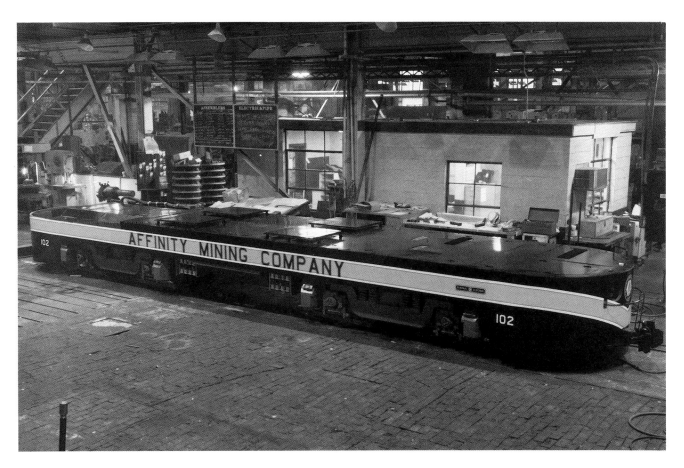

Affinity Mining Co. #102 2-Truck Electric Mine Locomotive (Trolley Lowered)

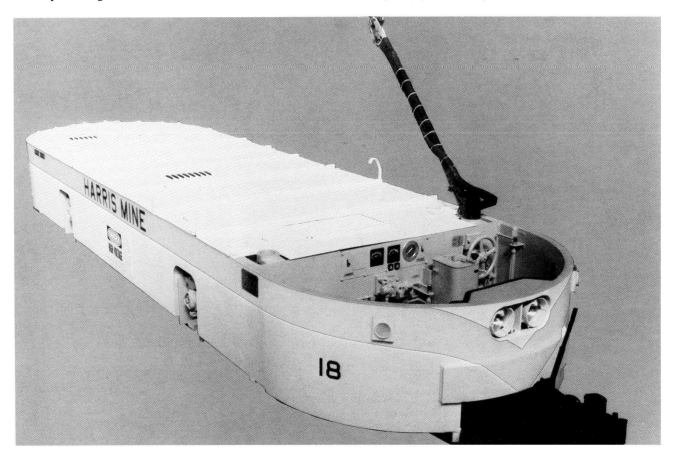

Harris Mine #18 Single Truck Electric Mine Locomotive (Trolley Raised)

COLOR SECTION

Locomotives identified by Operator, Model, HP. (15 Diesels, 1 Electric)

Atchison, Topeka & Santa Fe	#7400, Model B39-8, 3900hp
Burlington Northern	#5497, Model B39-8, 3900hp
Conrail	#5075, Model B32-8, 3170hp
Conrail	#6000, Model C39-8, 3900hp
Conrail	#6000, Model C39-8, 3900hp
General Electric	#505, Model C36-7, 3600hp
General Electric	#808, Model B39-8, 3900hp
General Electric	#8511, Model B39-8, 3900hp
Nacional De Mexico	#9339, Model C36-7, 3600hp
New York, Susquehanna & W.	#4002, Model Dash 8-40B, 4000hp
Norfolk Southern	#8612, Model C39-8, 3900hp
Southern (Norfolk Southern)	#3820, Model B36-7, 3600hp
Union Pacific	#9183, Model Dash 8-40C, 4000hp
Exxon Coker	#3, Model SL110, 110-ton Switching
People's Republic of China	#0305, Model C36-7, Rated 4000hp 1986
Nacional De Mexico	#EA001, Model E60C, 6000hp Passenger & Freight.

C36-7 Specifications

ratings, weights, dimensions

RATINGS
Continuous horsepower to generator for traction under standard conditions
with a General Electric 16-cylinder Model FDL16 engine .... 3600 hp
Continuous tractive effort with 83:20 gear ratio .... 96,900 lb.
Locomotive speed with 83:20 gear ratio, 40-inch wheels .... 70 mph
  (other gear ratios are available)

WEIGHTS
Minimum locomotive (fully loaded) .... 366,600 lb.
Per axle (fully loaded) .... 61,100 lb.
Locomotive weight subject to manufacturing tolerance of ± 2%
Modifications may increase weights.

WHEEL ARRANGEMENT .... C-C

MAJOR DIMENSIONS
Length inside knuckles .... 67 ft. 3 in.
Height .... 15 ft. 4½ in.
Width over handrails .... 10 ft. 2¼ in.
Minimum radius of curvature, locomotive alone .... 273 ft. (21°)

CAPACITIES
Fuel .... 3250 gal.
Engine lubricating oil .... 380 gal.
Cooling water .... 365 gal.
Sand .... 60 cu. ft.

performance features

New Series features incorporated in the 3600-hp, six-axle C36-7 are representative of General Electric's continuing design, evaluation, and manufacturing program to improve locomotive reliability, availability, and maintainability, and to reduce total operating costs.

The high performance capability of this model addresses the needs of the railroad industry for ever-increasing productivity.

DESIGN—The locomotive operating cab provides optimum visibility for operating in either direction. The control station is at the right side with the short hood leading.

POWER—A four-stroke-cycle, turbocharged diesel engine is the power source. An alternator, directly connected to the engine, furnishes power to axle-mounted traction motors. Full utilization of the engine's horsepower is available throughout the speed range of the locomotive.

OPERATION—A master controller as well as independent and automatic air brake valves are conveniently located to permit operation with either end leading. Direction of motion is controlled by a reverse lever. Throttle and reverse levers are interlocked to prevent operation of the reverser unless the throttle handle is in the off position.

SAFETY—Design of the locomotive reflects General Electric's continuing concern for safety. All safety appliances are in accordance with General Electric's interpretation of current FRA regulations.

TESTING—All General Electric locomotives are manufactured to stringent quality standards. Component parts are given standard commercial tests before assembly on the locomotive. Control wiring is checked by observing the sequence of contactor and relay operation and by testing for continuity of circuit between terminals. High-potential tests of traction and control circuits are made in accordance with current U.S.A. standards. Air brake tests assure proper operation. The power plant is tested at full load to check alternator and engine, including power and speed.

PAINTING—Durable finishes include interior gray enamel, black underframe and running gear, and special acid-resisting interior battery compartment paint. Exterior color and design are specified by the customer.

running gear

The running gear of the locomotive consists of two lateral motion swivel trucks. Center plate load is distributed by the cast-steel "floating bolster" to four rubber mounts which rest on the truck frame and provide controlled lateral motion. The cast-steel frame is supported by alloy steel coil springs over the journal boxes. Friction-type snubbers damp vertical oscillation.

WHEELS—Multiple-wear, rolled steel wheels with AAR tread and flange contour are furnished.

AXLES—Forged steel axles, conforming to AAR material specifications are provided.

JOURNALS—Journals are equipped with sealed grease-lubricated roller bearings. Pedestal openings on the truck frame have renewable non-metallic wear plates.

CENTER PLATES—Equipped with non-metallic liners, and protected by dust guards.

SAFETY HOOKS—Minimize slewing in case of derailment and permit trucks to be lifted with the locomotive superstructure.

locomotive brakes

AIR BRAKES—Schedule 26L air brake equipment with 26F control valve is furnished. The air brakes may be operated independently or with train brakes. Connections for furnishing compressed air to the train brakes are provided at each end of the locomotive.

BREAK-IN-TWO PROTECTION—Prevents the possible release of brakes from an emergency application initiated in the train with the brake valve handle in its release position.

COMPRESSOR—One 3-cylinder, 2-stage water-cooled engine-driven air compressor furnishes air for the locomotive and train braking systems.

RESERVOIRS—Reservoir capacity of 56,000 cu. in. is furnished for storing and cooling air for the brake system. Both main reservoirs are equipped with automatic drain valves.

BRAKE EQUIPMENT—Brake cylinders are supported by the truck frames and operate fully equalized brake rigging. Rigging is furnished with hardened steel bushings and adjustment to compensate for wheel and shoe wear.

HAND BRAKE—Is located on the outside of the short hood and provided to hold the locomotive at standstill.

DYNAMIC BRAKING—Brakes the locomotive electrically using the traction motors as generators and dissipating the electric power in resistors. Interlock is included to prevent application of air brakes on a locomotive in dynamic braking when automatic air is applied to the train. With standard GE braking, engine speed is automatically matched to the cooling requirements as dictated by braking level.

underframe

The welded underframe is made of rolled steel sections and plate. Hoods, cabs, equipment, and tanks are supported by the main frame members. Space between these members is enclosed to form an air duct which distributes clean air throughout the locomotive.

WEARPLATES—Renewable, wear-resistant steel plates are applied to side bearing pads, and draft gear housing.

COUPLERS—AAR type E top-operated couplers with NC-391 rubber-cushioned draft gear and alignment control are provided at each locomotive end.

PILOT AND SIDE STEPS—A pilot is at each end of the locomotive. Side steps provide access to the platform.

LIFTING AND JACKING—Four jacking pads in combination with lugs for cable slings are provided on the side bolsters.

FUEL TANK—A heavy gage welded steel tank is bolted to the underframe between the trucks. The tank is provided with baffle plates, clean-out plug, water drains, and vent. Filler connections and fuel level gages are furnished on each side of the locomotive. Emergency fuel shutoffs are provided.

superstructure

The welded steel superstructure consists of a short front hood, operator's cab, engine hood, and radiator compartment. Engine hood is bolted to the underframe and is removable.

SHORT HOOD—The short hood contains a top-serviced sandbox. A door in the front bulkhead of the cab provides access. Classification lights are mounted on this hood. A ventilator is provided.

OPERATOR'S CAB—Sides and roof are insulated and steel-lined. The floor, raised above the underframe, is covered with high density hardboard. The cab has windows in the front and rear. Two-pane center windows on each side have sliding sash equipped with latches. Doors in diagonally opposite corners of the cab provide access to walkways on both sides of the locomotive. They have windows, weather stripping, and locks. All cab glazing is certified in compliance with FRA safety glazing standards. Headlights and number boxes are arranged on the outside above the front windows. Electric cab heat is provided.

WALKWAYS—Walkways with handrails and non-skid treads are at each end of the locomotive and along the hoods.

ENGINE COMPARTMENT—Encloses the engine, blower, and traction alternator. Complete access to this equipment is provided by doors the full height and length of the hood extending along both sides of the locomotive. Doors in the roof provide overhead access to cylinders. Detachable roof sections permit removal of equipment.

RADIATOR COMPARTMENT—Contains the radiators, fan and gearbox, compressor, and engine air cleaners. The compressor is enclosed in the radiator compartment with free air access from the engine compartment. The radiators are roof-mounted. A reinforced screen over the air outlet opening is removable to permit removal of the radiators, fan, and gearbox. Dynamic braking grids are mounted along each side of the radiator compartment. An end section holds a sandbox, serviced from the roof. Rear headlights, classification lights, and number boxes are mounted on this section.

EQUIPMENT COMPARTMENTS—Main propulsion control equipment is located on the left side of the locomotive beneath the operator's cab. This compartment, maintained under positive air pressure to keep out dirt and water, contains contactors, reverser, and braking switch. Excitation and other panels and devices are located in a compartment behind the operator's cab. The compartment is gasketed to prevent entrance of dust. It can be accessed from either walkway but is not accessible from the operator's cab. Air brake devices are located in a compartment along the right side of the locomotive under the operator's cab. Battery trays are located in a box with hinged top doors for ease of servicing.

VENTILATION—Filtered air is provided through self-cleaning air cleaners located in the underframe. Clean air is delivered under pressure for equipment cooling and pressurization, and cab ventilation. Engine air is cleaned by self-cleaning air cleaners and by General Electric paper filters.

location of major systems

1. Engine - GE Model 7FDL16
2. Alternator
3. Auxiliary Generator
4. Rectifiers
5. Equipment Blower
6. Air Compressor
7. Radiator Fan Gear Unit
8. Engine Exhaust Stack
9. Engine Air Filters
10. Engine Water Tank
11. Lube Oil Cooler
12. Lube Oil Filter
13. Radiator
14. Braking Resistors
15. Sand Box
16. Number Box
17. Sand Fill
18. Fluid Amplifier
19. Battery Box
20. Upper Control Compartment
21. Lower Control Compartment
22. Fuel Tank
23. Fuel Filler
24. Toilet
25. Engine Control Panel
26. Battery Switch
27. Control Console
28. Air Brake Valve
29. Electric Cab Heater
30. Side Strip Heaters
31. Sliding Seats
32. Hand Brake
33. Equipment Air Filters
34. Air Duct
35. Eddy Current Clutch

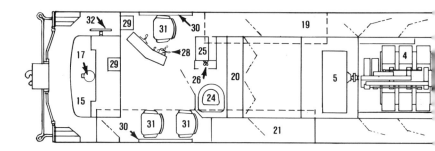

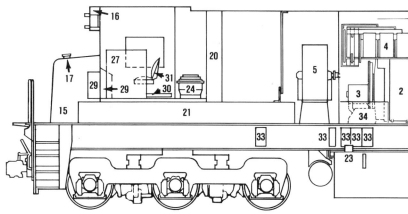

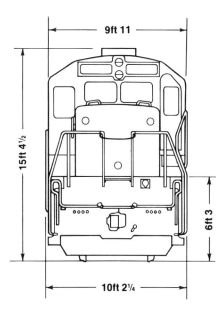

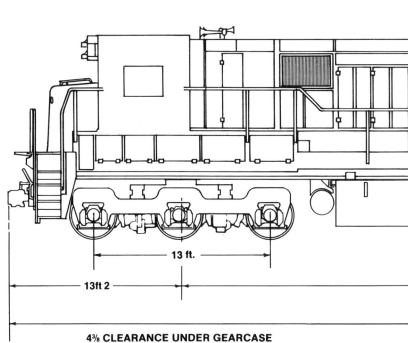

4⅜ CLEARANCE UNDER GEARCASE

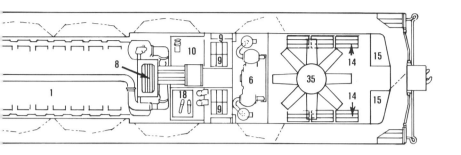
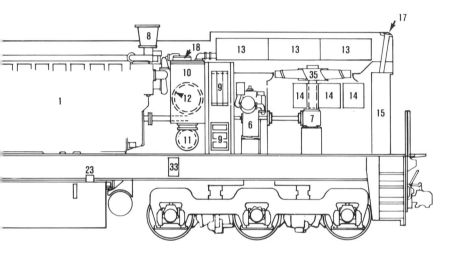
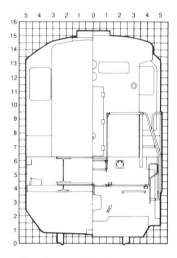

Maximum Equipment Diagram

Left half-section through exhaust stack.
Right half-section facing rear end of locomotive.
Maximum tolerance on height: ± 1½ inches.

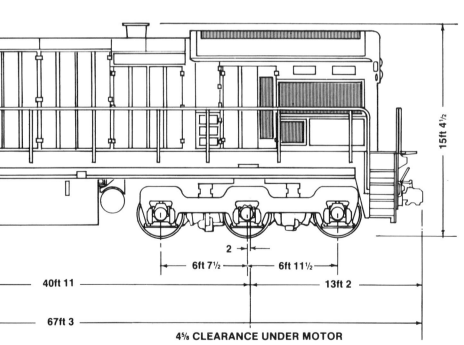

power plant

DIESEL ENGINE—Type—one General Electric FDL; cylinder arrangement—45°V; stroke cycle—4; bore and stroke—9 X 10½ inches; RPM—1050; turbocharger—one.

GOVERNOR—Self-contained, electro-hydraulic Woodward PG type governor automatically regulates horsepower output at each throttle setting.

OVERSPEED PROTECTION—The engine automatically shuts down if speed exceeds maximum rated rpm by 10%.

COOLING SYSTEM—A gear-driven centrifugal pump integral with the diesel engine circulates cooling water through the engine, turbocharger, air intercoolers, self-draining radiators, lubricating oil cooler and air compressor. The tank is equipped with a sight gage to indicate water level and with screens which provide maximum filtration of debris and scale. The system is pressurized. Abnormally low water pressure automatically shuts down the engine.

TEMPERATURE CONTROL—A solid-state fluid amplifier control system and variable speed radiator fan automatically maintain cooling system temperature. The fluid amplifier regulates the flow of cooling water through the radiator sections. The radiator fan eddy current clutch matches fan speed/horsepower to cooling requirements.

FUEL SYSTEM—A motor-driven pump transfers fuel from the tank through a strainer and filter to the injection pumps. Each cylinder is equipped with a high pressure fuel injection pump and injector.

LUBRICATING SYSTEM—A single pressure-regulated system is supplied by a gear type pump integral with the diesel engine. A lubricating oil reservoir is located in the engine subbase. Suction strainer, lubricating oil filters, and water-cooled oil cooler are provided. Abnormally low lubricating oil pressure or abnormally high crankcase pressure automatically shuts down the engine.

ENGINE STARTING—The engine is cranked by the two GY27's from storage battery power.

HORSEPOWER OUTPUT—Horsepower input to the alternator for traction is provided under AAR standard conditions with specified fuel and lubricating oil.

LOW IDLE—Has been provided to reduce engine speed during prolonged periods of idle operation.

electric transmission

TRACTION MOTORS—GE-752AF traction motors are furnished. They are direct current, series wound, and separately ventilated by the clean air system. The armatures are mounted in anti-friction bearings. Motors drive through single-reduction spur gearing. They are supported by the axles to which they are geared and by resilient nose suspensions on truck transoms.

TRACTION ALTERNATOR—One General Electric GTA-24 traction alternator is mounted directly on the engine. It is an alternating current, single anti-friction bearing, separately-excited machine. The output is rectified by a full wave rectifier.

CONTROL—Railway type single-end control is provided. Control devices are grouped in two steel compartments, fitted with access doors. Reverser, braking switch and line contactors are electro-pneumatically operated. Other contactors are magnetically operated. Circuit breaker-type switches are used in control circuits where overcurrent protection is required. Sanding is train-lined automatically. Alternator transition is automatic.

EXCITER AND BATTERY-CHARGING GENERATOR—Two Model GY27 exciters are gear-driven from the traction alternator. One provides controlled excitation of the alternator field. The other furnishes power at regulated potential for battery-charging, lighting and control.

STORAGE BATTERY—A 32-cell lead acid type, 420 ampere hour battery starts the engine and furnishes power for lights and other auxiliaries when the engine is shut down.

WHEELSLIP CORRECTION—The SENTRY Adhesion Wheelslip System uses comparative motor speed as a means of detecting slips. Highly sensitive inductive speed sensors are indexed on a special gear mounted on the motor shaft. Slip is corrected by automatic application of sand and reduction of power. Compensation for wheel size variations is automatic.

GROUND RELAY PROTECTION—If a ground occurs, engine speed returns to idle, power is removed, the alarm bell rings, and visual indication is given to the operator.

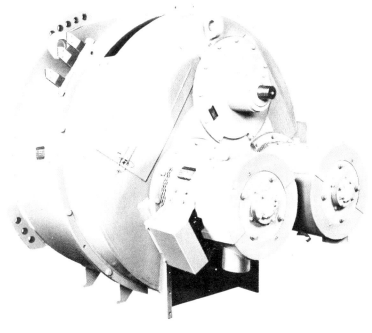

operating controls

Controls and instruments are grouped at the operator's station and auxiliary panels in the operator's cab. The controller/console meets all AAR requirements.

OPERATING CONTROLS
Controller with throttle, dynamic braking
   handle and reverser-selector levers
Engine start push button
Engine stop push button and emergency fuel shutoff
Brake valves
Bell ringer valve
Air horn valve
Window wiper valves
Circuit breakers and switches
Emergency multiple unit engine stop switch
Sander switch
Lead axle sander switch
Ground relay reset push button
Engine control switch

INSTRUMENTS
Two brake gages with test fittings
Dual reading load meter
Mechanical-type speed recorder and odometer
Fuel oil pressure gage.
WARNING INDICATORS
Warning light—wheelslip, PCS,
   annunciator light
Alarm bell and warning light
   High crankcase pressure
   Ground relay
   Engine overtemperature
   No battery charge
   Governor shutdown
   Engine air filters
   Hot diodes

locomotive accessories

BELL—Cast iron with air-operated ringer and operating valve.
CAB HEAT—Electric.
CLASSIFICATION LIGHTS—Two, 3-aspect electric lights at each end of the locomotive.
CLOTHES HOOKS—Provided in operator's cab.
CONDUCTOR'S EMERGENCY VALVE—Standard AAR location.
EMERGENCY FUEL SHUTOFF—Three pushbuttons, one on each side of the underframe and one in the operator's cab.
FIRE EXTINGUISHERS—Two, 20-pound dry chemical, one at each end of the locomotive.
FUEL GAGES—One on each side and one on each end for upper and lower level readings.
HEADLIGHTS—At each end of the locomotive to comply with FRA requirements. Dimming control provided with the lights.
HORN—One, 3-tone with two bells forward and one to the rear.
INTERIOR LIGHTS—For illuminating the operator's cab, nose cab, and instruments.
SANDERS—Eight, pneumatically operated, to sand ahead of the lead wheels of each truck in each direction.
SEATS—Two, wall-mounted, adjustable for height and for operating in either direction. Cushioned arm rests provided at side windows.
STEP LIGHTS—Four, one for each side step.
SUN VISOR—Adjustable-type.
WATER TEMPERATURE GAGE—Located in the engine compartment.
WINDOW WIPERS—Six, air-operated, on front and rear windows of operator's cab.

modifications

The following modifications may increase locomotive weight, dimensions and price.

ADDITIONAL FUEL—750 gallons for a maximum fuel capacity of 4000 gallons.
AUTOMATIC SANDING—Sanding in either direction initiated in the event of emergency brake application.
AWNINGS—Over window on each side of cab.
BALLAST—Locomotive can be ballasted up to 420,000 pounds.
BATTERY CHARGING AMMETER—Mounted on engine control panel.
BATTERY CHARGING RECEPTACLE—One 150-ampere type.
BRAKE SHOES—To meet customer requirements.
CAB SIGNAL EQUIPMENT—Train control cab signal or train speed control equipment as now used on various railroads.
COUPLER—AAR Type F in place of Type E. NC-390 draft gear required.
DELUXE CAB SEATS—Upholstered with armrests.
DRAFT GEAR—NC-390 equipment furnished.
DYNAMIC BRAKING—Furnished as standard equipment.
EXTENDED RANGE DYNAMIC BRAKE—Equipment provided with dynamic braking to obtain greater braking effort at low train speed.
EXTRA CAB SEAT—Third seat in operator's cab.
FIRE EXTINGUISHER—To meet customer requirements.
FUEL LEVEL GAGES—Dial-type gages provided on both sides of the tank near the filler openings.
GEAR RATIOS—Optional gear ratios available as follows:

Gears	81:22	80:23	79:24
Ratio	3.68	3.48	3.29
Max. mph	75	79	84

LONG HOOD LEADING CONTROL—Control station mounted fo operation with the long hood leading
MULTIPLE-UNIT CONTROL—Furnished as basic for operating two or more units with 26L or 24RL air brakes from one cab.
ON-BOARD LOAD TESTING—To permit loading for test purposes on the dynamic braking grids.
OVERSPEED PROTECTION—To return engine to idle, cut off power, and make automatic brake application.
RETENTION TANK—To hold liquids collected by platform drains.
SAFETY CONTROL—Safety deadman control including foot pedal valve, time delay, warning whistle, and brake application.
TOILET—Electric incinerating, biodegradable, or retention type. Located in compartment at the rear of the operator's cab.
TOOL BOX
TRAIN COMMUNICATION—As used by various railroads.
TRUCK JOURNAL BEARINGS—Hyatt bearings in place of the standard Timken GG bearings.
TWO-STATION CONTROL—For operating the locomotive from either of two diagonally opposite positions in the operator's cab.
VISUAL WARNING SIGNAL LIGHTS—Located at each end of the locomotive.
WALKWAYS—To permit passage between units.
WATER COOLER/REFRIGERATOR—Floor-mounted in operator's cab.
WINDSHIELD WINGS—Wind deflectors; one in front and rear of each side window.

performance characteristics

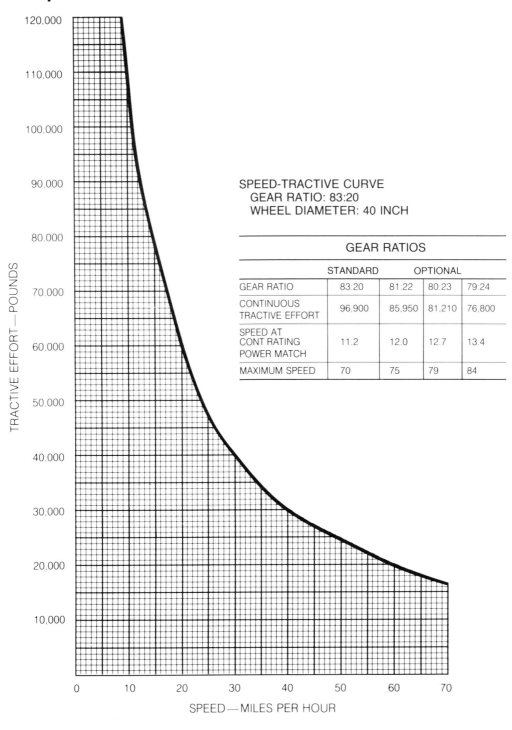

SPEED-TRACTIVE CURVE
GEAR RATIO: 83:20
WHEEL DIAMETER: 40 INCH

GEAR RATIOS				
	STANDARD	OPTIONAL		
GEAR RATIO	83:20	81:22	80:23	79:24
CONTINUOUS TRACTIVE EFFORT	96,900	85,950	81,210	76,800
SPEED AT CONT RATING POWER MATCH	11.2	12.0	12.7	13.4
MAXIMUM SPEED	70	75	79	84

for increased motive power productivity.

To fully utilize this increased traction motor capacity, the C36-7 provides dramatic improvements in adhesion through the new SENTRY Adhesion Control System.

To achieve tractive effort ratings compatible with increased productivity, General Electric utilizes a new finer pitch, 70-mph (83/20) gearing designed to minimize wear and reduce vibration by providing more surface area and longer contact for mating teeth.

The new GE-752AF traction motor with the 83/20 gear ratio provides a continuous tractive effort rating of 96,900 pounds per locomotive. One of the most significant electrical system advances incorporated into the C36-7 is the SENTRY Adhesion Control System. As applied on the locomotive, this system is capable of taking full advantage of the higher continuous and short-time motor ratings and provides marked improvements in all-weather adhesion.

TEST RESULTS WET RAIL WITH AUTOMATIC SANDING

C36-7 WITH SENTRY ADHESION SYSTEM

CONVENTIONAL DISPATCH ADHESION

Innovations in the GE-752AF traction motor significantly increase ratings.

The C36-7: continuing fuel

Fuel economy has received major attention on the C36-7 as it has on all New Series models.

The C36-7 fuel economy package includes these features:

- The General Electric four-stroke-cycle diesel engine
- General Electric 1616B4 turbocharger
- Optimum engine speed schedule
- Dynamic braking – modulated grid cooling
- Low engine idle
- Increased notch horsepower
- Fuel injection improvements
- Variable speed radiator fan

Four-stroke-cycle diesel engine

The inherent fuel efficiency of General Electric's four-stroke-cycle diesel engine has been enhanced over the past 20 years through improvements in the fuel system, turbocharger, pistons and liners, intercooler, and exhaust system. General Electric believes that the present configuration represents the most advanced state-of-the-art in fuel efficiency of railroad locomotive diesel engines.

General Electric 1616B4 turbocharger

While the principal objective of the General Electric turbocharger development program was to improve reliability and performance, major advances in fuel efficiency have also been achieved. Particular emphasis was placed on increasing the operating range in altitude and ambient temperature.

The B4 model of the General Electric turbo reaches another plateau in the continuing program for fuel efficiency by improving the turbocharger seal system.

The General Electric diesel engine represents advanced state-of-the-art design concepts.

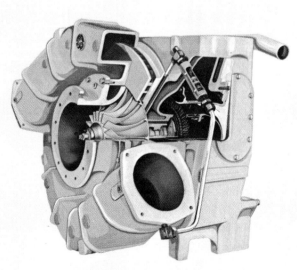

General Electric turbocharger provides reliability, performance, and fuel efficiency.

Reduction of compressed air to the turbocharger air seal

This improved turbocharger seal system has reduced the amount of compressed air required to prevent carbon build-up and consequent seal failure. The resultant reduction in the air compressor duty cycle further contributes to system fuel efficiency.

Optimum engine speed schedule

Turbochargers applied to early 3600 horsepower locomotives developed excessive pre-turbine temperatures under some operating conditions. To compensate for this condition, a modified engine speed schedule allowed the engine to accelerate to higher speed while the traction load was relatively low. For example, notches 6, 7, and 8 had a notch 8 speed (1050 rpm). As all the auxiliary load is mechanically driven, the higher speeds added to load and fuel consumption.

In the C36-7, the lower operating temperature of the General Electric turbocharger obviates the need for a modified speed schedule. The standard schedule has thus been applied, significantly reducing auxiliary load and fuel consumption.

Dynamic braking – modulated grid cooling

General Electric locomotives utilize the radiator fan to cool dynamic grids. In the past, notch 8 speed was maintained regardless of the braking level. While this achieved a degree of simplicity, it did not address today's fuel concerns.

The system has been modified to automatically match engine speed to cooling requirements as dictated by braking level. The average dynamic braking fuel rate is substantially reduced.

efficiency for the years ahead.

Low engine idle

A low idle speed feature is provided to reduce fuel consumption during the significant portion of idle time in most duty cycles. A temperature switch has been added to prevent low idle operation if water temperature drops below an acceptable minimum level.

Increased notch horsepower

On the C36-7, gross horsepower is maintained constant by transferring reductions in auxiliary horsepower to horsepower for traction. As useful horsepower becomes a higher percentage of total horsepower, fuel efficiency, measured by alternator output is improved. The resulting constant rack setting benefits maintenance personnel.

Fuel injection system improvements

A number of new fuel injection system improvements are incorporated into the C36-7 locomotive.

The injection nozzle spray angle has been increased from 150 to 157 degrees. The new spray angle permits better fuel and air mixing, and thus provides more efficient combustion and significant fuel savings.

Increased spray angle permits better fuel and air mixing.

Electrostatic discharge machined fuel injection tips produce a finer spray for improved fuel/air mixing. This is a result of improved manufacturing tolerances.

New larger 18-mm double helix fuel injection pumps inject fuel within a shorter period of time and vary the timing at different throttle notches. This optimizes fuel injection timing at lower throttle notches and provides more efficient combustion throughout the horsepower range of the locomotive engine.

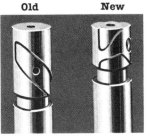

Larger double helix pump provides more efficient combustion.

It is estimated that the above fuel injection modifications will improve fuel efficiency an additional 1.5% when applied to a typical freight duty cycle.

Variable speed radiator fan

This hub-mounted radiator fan clutch reduces the average auxiliary load on the engine when radiator cooling requirements are less than maximum. The fan has three speeds – off, low, and full.

In notch 8, the reduction in auxiliary horsepower is applied as horsepower for traction when the fan is in low speed or off.

Fuel savings from this feature are estimated at 2.0%.

Estimated saving

The fuel economy features described make up the 1980 and 1981 fuel economy packages applied to New Series locomotives. Utilizing a typical freight duty cycle, it is possible to estimate the fuel savings these two packages provide on a C36-7 as compared to an earlier U36C.

High-efficiency variable speed radiator fan contributes to improved fuel economy.

The 1980 package provides a 7.5% savings: the 1981 package, a 3.9% savings. In addition to the consideration of reduced operating cost, this level of fuel efficiency offers a counter thrust to the prospect of an adverse impact on rail revenues caused by limited fuel availability.

Returning to the economic evaluation, the following graph allows various estimates of average future fuel prices to be tested against the present worth of savings. Each railroad can assess its estimate of the true present value of the C36-7 fuel economy. For example, the combined 11.4% savings of the 1980 and 1981 fuel packages and an average price of $1.00 per gallon over 15 years represent a present worth of $279,900 per locomotive.

All railroads do not operate on the typical duty cycle and may burn more or less fuel per high horsepower unit. The present worth savings are directly proportional to fuel usage. Your General Electric representative can provide an analysis for your particular railroad.

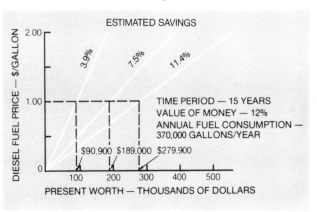

The C36-7: reaching a new milepost

The 16-cylinder, four-stroke-cycle engine on the C36-7 results from two decades of General Electric experience in U.S. railroad diesel technology. This mature design represents the most up-to-date advances in that technology.

General Electric turbocharger

General Electric designed and built turbochargers, first introduced in 1976, significantly reduce operating temperatures, while improving component reliability and durability.

The General Electric 1616B4 turbo furnished on the C36-7 maintains these improvements while achieving even further gains in fuel economy.

The ability to maintain full horsepower over the complete range of altitude and ambient temperature eliminates the need for automatic deration.

General Electric turbochargers significantly reduce operating temperatures and improve component durability.

Steel crown pistons

Early experience at 3600 horsepower demonstrated the need for piston improvement. The steel crown piston was introduced to provide the high strength and heat resistance required in the crown area. Added wear resistance is furnished in the ring area.

Since its introduction in 1972, the steel crown design has proven its high reliability and long-life capability.

Steel crown piston provides excellent heat and wear resistance.

in diesel engine technologies.

Main and connecting rod bearings

The grooveless main bearing has demonstrated a reduced wear rate and is expected to provide a 50 to 100% increase in service life over earlier designs.

Elimination of the oil groove in the lower main bearing results in a thicker oil film and lower unit area loadings.

Introduced in mid-1976, the grooveless design has proven to be more tolerant of microscopic dirt particles and degraded oil conditions.

The successful experience gained with grooveless main bearings has led General Electric to apply this principle to its conncecting rod bearings. This combined with non-destructive testing procedures that accurately control bearing thickness has resulted in improved durability and reliability.

Improvements in bearing design extend service life.

Intake and exhaust valves

Since mid-1974, 15° inlet valves have replaced 45° valves on General Electric diesel engine cylinders. These 15° one-piece silchrome intake valves have a smaller angle that reduces scrubbing action between the valve and seal, minimizing wear. As wear was predominately on the valve, longer service life has resulted. The exhaust valves are two-piece with Inconel 751 heads, fully shot peened and chrome plated.

15° inlet valves reduce scrubbing action between valve and seat, minimizing wear.

Cylinder liner improvements

General Electric has changed the process used to harden its engine cylinder liners. The new, patented, Melonited procedure was adopted to meet disposal requirements of current environmental laws. Disposal of materials remaining from this process causes far fewer problems than the previously-used Tufftride, also patented, procedure. Standards of hardness, hardness depth, and resulting preformance have not changed. Several years of field testing preceded the decision to adopt the new process.

Hardened liners resist severe conditions of temperature and lubrication.

New lube oil pump

Tapered roller bearings, applied to the lube oil pump, enhance reliability, provide longer service life and offer heavier load carrying capabilities than previously used bearings.

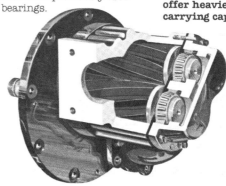

New overspeed system

A new overspeed system with a layshaft is now being applied to New Series locomotives. Designed for improved availability and ease of engine starting, this new system overcomes the shortcomings of previous systems. It provides the layshaft for ease of starting. And, the new system also incorporates a preset overspeed link. This eliminates waiting for overspeed governor oil pressure to build prior to start-up of the engine. This feature in conjunction with the start station located at the governor eliminates the need for additional persons to start an engine during adverse weather conditions.

New overspeed system improves availability and eases engine starting.

Tapered roller bearings offer heavier load carrying capability.

-95-

The C36-7: enhancing 3600 horsepower

SENTRY adhesion control

One of the most significant advances provided in the C36-7 is the SENTRY Adhesion Control System.

Objective of the design effort that led to this development was providing higher levels of all-weather adhesion without compromising wheel and rail life or sacrificing fuel efficiency.

The SENTRY Adhesion Control System has been tested under adverse conditions of wet rail, severe grades, and reverse curves at adhesion levels from 23% to 26%.

These test results show that railroads can safely increase, on a regular basis, the tonnage assigned to the C36-7.

High resolution speed sensors

The SENTRY Adhesion Control System uses comparative axle speed as a means of detecting slips. Inductive speed sensor pick-ups read the teeth of a special gear, internally-mounted on the motor shaft.

The motor-mounted sensor is approximately six times more sensitive than the axle-mounted alternator on earlier General Electric locomotives. It is located to avoid damage.

Automatic compensation for wheel size variation

In the past, the effectiveness of wheelslip detection and control systems has been limited by the need for the systems to accommodate variations in wheel diameter.

The SENTRY Adhesion Control System overcomes this problem by calibrating the circuitry for relative size of the various wheels. This calibration is accomplished automatically each time the locomotive is placed in a "coast" mode.

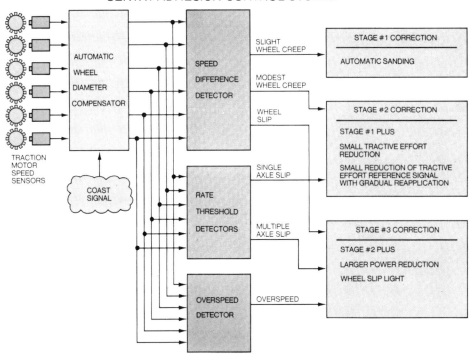

Full sensitivity of the system is utilized, thus providing the earliest detection of slips.

Control of slips

The SENTRY detection system is extremely sensitive to early slips. Initially, a first-stage correction is applied. This results in an automatic application of sand. The first-stage system controls the vast majority of slips without the need for a power reduction.

A second-stage correction is automatically employed if necessary. This includes automatic sanding as well as a small horsepower reduction to decrease tractive effort followed by gradual power reapplication. A third-stage system utilizes all second-stage corrections and a larger power reduction. The system automatically returns to the previous power level as conditions improve.

Parallel operation of motors

With a parallel motor connection, a slipping motor does not directly affect the performance of a second motor, as is the case with series-parallel operation. Through adoption of parallel motors alone, adhesion is inherently improved.

GE-752AF traction motor

The C36-7's advanced SENTRY Adhesion Control System allows full utilization of the higher continuous current rating of the GE-752AF traction motor.

The GE-752AF electrical design incorporates improvements that have been successfully applied to other General Electric traction motors over the past several years.

These improvements, which included complete Class H motor insulation, produced a 6% increase in motor rating.

Armature coils utilize two conductors in parallel rather than a single conductor, resulting in a reduction in eddy current losses.

The armature coil insulation system has been improved. This has permitted a 7.4% increase in the cross-section area of copper in the armature coils.

The symmetrical-radial-tangential armature winding design and the reaction brushholder provide better commutation in both directions of armature rotation. No compromise has been made in ease of rewinding the armature.

Field coils utilize an upgraded insulation system for moisture resistance and improved high temperature performance.

operation through electrical developments.

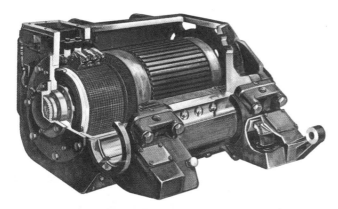

GE-752AF motor increases locomotive continuous tractive effort rating.

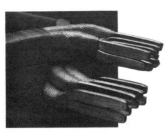

With conductors in parallel, eddy current losses are reduced.

Improved armature winding design provides better commutation.

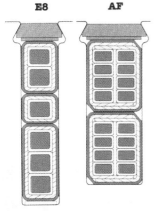

Comparison of E8 and AF armature coils illustrates a 7.4% increase in copper cross-section.

Reaction brushholder also contributes to improved commutation.

GTA-24 alternator and rectifier

For almost a decade, General Electric alternators have been successfully applied at 3600 horsepower. Building on this experience, the GTA-24 has been developed to provide the high current output required for parallel motor connections. This is part of the overall thrust for higher six-axle locomotive performance.

The traction motors of the C36-7 are permanently connected in parallel to complement the SENTRY Adhesion Control System in providing improved six-axle adhesion.

To provide the higher starting current inherent in six-parallel start as well as the high voltage needed for full horsepower utilization at maximum locomotive speeds, the alternator has been equipped with two windings.

When high starting current is required, the windings are connected in parallel. This provides twice the current available from a single winding.

For high-speed operation, the windings are connected in series. This offers the higher voltage needed for full horsepower utilization throughout the full locomotive speed range. Reconnection of the winding from parallel to series is automatic.

To meet the increased demand for starting conditions, the current capability of power rectifiers has been increased.

Improved power rectifier arrangement increases current capacity.

PARALLEL CONNECTION FOR HIGH CURRENT, LOW SPEED

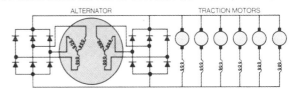

SERIES CONNECTION FOR HIGH VOLTAGE, HIGH SPEED

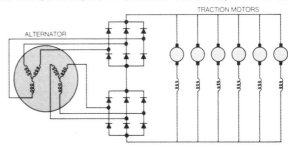

The C36-7: mechanical system refinements

Major mechanical improvements center on application of a new gearcase design, 83/20 gearing, and greaseless couplings.

New gearcase design

Smooth running 83/20 gearing is housed in a new gearcase. Stiffness has been increased to allow the use of machined and gasketed joints between top and bottom halves. This eliminates the need to maintain top and bottom halves as matched sets.

Machined joints allow use of more efficient seals in the bore area and result in reduced lubricant migration. The new gearcase design features an improved bolting arrangement with mounting points on the bottom half only. This permits use of standard hardware and prevents distortion when attaching to the motor.

New gearcase design results in reduced leakage and simplified maintenance.

New greaseless couplings

Non-lube Dynaflex® couplings for the equipment air blower, air compressor, and radiator fan drives eliminate periodic lubricating requirements. These couplings absorb torsional shock and reduce drive train stresses, thus providing longer life and lower life cycle costs.

Applied to equipment air blower, air compressor, and radiator fan drive, the greaseless couplings lengthen service life.

Non-lube Dynaflex coupling eliminates periodic lubrication requirements.

Greaseless coupling absorbs torsional shock and reduces drive train stresses.

C30-7A Specifications

ratings, weights, dimensions

RATINGS
Continuous horsepower to generator for traction under standard conditions
with a General Electric 12-cylinder Model FDL 12 engine . . . . 3000 hp
Continuous tractive effort with 83:20 gear ratio 96,900
Locomotive speed with 83:20 gear ratio, 40-inch wheels . . . . 70 mph
  (other gear ratios are available)

WEIGHTS
Minimum locomotive (fully loaded) . . . . 359,000 lbs.
Per axle (fully loaded) . . . . 59,830 lbs.
Locomotive weight subject to manufacturing tolerance of ± 2%
Modifications may increase weights.

WHEEL ARRANGEMENT . . . . C-C

MAJOR DIMENSIONS
Length inside knuckles . . . . 67 ft. 3 in.
Height . . . . 15 ft. 4½ in.
Width over handrails . . . . 10 ft. 2¼ in.
Minimum radius of curvature, locomotive alone . . . . 273 ft. (21°)

CAPACITIES
Fuel . . . . 3250 gal.
Engine lubricating oil . . . . 300 gal.
Cooling water . . . . 350 gal.
Sand . . . . 60 cu. ft.

performance features

New Series features incorporated in the 3000-hp, six-axle C30-7A are representative of General Electric's continuing design, evaluation, and manufacturing program to improve locomotive reliability, availability, and maintainability, and to reduce total operating costs.

The high performance capability of this model addresses the needs of the railroad industry for ever-increasing productivity.

DESIGN—The locomotive operating cab provides optimum visibility for operating in either direction. The control station is at the right side with the short hood leading.

POWER—A four-stroke-cycle, turbocharged diesel engine is the power source. An alternator, directly connected to the engine, furnishes power to axle-mounted traction motors. Full utilization of the engine's horsepower is available throughout the speed range of the locomotive.

OPERATION—A master controller as well as independent and automatic air brake valves are conveniently located to permit operation with either end leading. Direction of motion is controlled by a reverse lever. Throttle and reverse levers are interlocked to prevent operation of the reverser unless the throttle handle is in the off position.

SAFETY—Design of the locomotive reflects General Electric's continuing concern for safety. All safety appliances are in accordance with General Electric's interpretation of current FRA regulations.

TESTING—All General Electric locomotives are manufactured to stringent quality standards. Component parts are given standard commercial tests before assembly on the locomotive. Control wiring is checked by observing the sequence of contactor and relay operation and by testing for continuity of circuit between terminals. High-potential tests of traction and control circuits are made in accordance with current U.S.A. standards. Air brake tests assure proper operation. The power plant is tested at full load to check alternator and engine, including power and speed.

PAINTING—Durable finishes include interior gray enamel, black underframe and running gear, and special acid-resisting interior battery compartment paint. Exterior color and design should be specified by the customer.

running gear

The running gear of the locomotive consists of two lateral motion swivel trucks. Center plate load is distributed by the cast-steel "floating bolster" to four rubber mounts which rest on the truck frame and provide controlled lateral motion. The cast-steel frame is supported by alloy steel coil springs over the journal boxes. Friction-type snubbers damp vertical oscillation.

WHEELS—Multiple-wear, rolled steel wheels with AAR tread and flange contour are provided.

AXLES—Forged steel axles, conforming to AAR material specifications are furnished as standard equipment.

JOURNALS—Journals are equipped with sealed grease-lubricated roller bearings. Pedestal openings on the truck frame have renewable non-metallic wear plates.

CENTER PLATES—Equipped with non-metallic liners, and protected by dust guards.

SAFETY HOOKS—Minimize slewing in case of derailment and permit trucks to be lifted with the locomotive superstructure.

DIESEL ENGINE —Type—one General Electric FDL; cylinder arrangement—45°V; stroke cycle—4; bore and stroke—9 x 10½ inches; RPM—1050; turbocharger—one.

GOVERNOR—Self-contained, electro-hydraulic Woodward PG type governor automatically regulates horsepower output at each throttle setting.

OVERSPEED PROTECTION—The engine automatically shuts down if speed exceeds maximum rated rpm by 10%.

COOLING SYSTEM—A gear-driven centrifugal pump integral with the diesel engine circulates cooling water through the engine, turbocharger, air intercoolers, self-draining radiators, lubricating oil cooler and air compressor. The tank is equipped with a sight gage to indicate water level and with screens which provide maximum filtration of debris and scale. The system is pressurized. Abnormally low water pressure automatically shuts down the engine.

TEMPERATURE CONTROL—A solid-state fluid amplifier control system and variable speed radiator fan automatically maintain cooling water temperature. The fluid amplifier regulates the flow of cooling water through the radiator sections. The radiator fan eddy current clutch matches fan speed/horsepower to cooling system requirements.

FUEL SYSTEM — A motor-driven pump transfers fuel from the tank through a strainer and filter to the injection pumps. Each cylinder is equipped with a high pressure fuel injection pump and injector.

LUBRICATING SYSTEM—A single pressure-regulated system is supplied by a gear type pump integral with the diesel engine. A lubricating oil reservoir is located in the engine sub-base. Suction strainer, lubricating oil filters, and water-cooled oil cooler are provided. Abnormally low lubricating oil pressure or abnormally high crankcase pressure automatically shuts down the engine.

ENGINE STARTING—The engine is cranked by the two GY27's from storage battery power.

HORSEPOWER OUTPUT —Horsepower input to the alternator for traction is provided under AAR standard conditions with specified fuel and lubricating oil.

LOW IDLE—Has been provided to reduce engine speed during prolonged periods of idle operation.

location of major systems

1. Engine – GE Model 7FDL12
2. Alternator
3. Auxiliary Generator
4. Rectifiers
5. Equipment Blower
6. Air Compressor
7. Radiator Fan Gear Unit
8. Engine Exhaust Stack
9. Engine Air Filters
10. Engine Water Tank
11. Lube Oil Cooler
12. Lube Oil Filter
13. Radiator
14. Braking Resistors
15. Sand Box
16. Number Box
17. Sand Fill
18. Fluid Amplifier
19. Battery Box
20. Upper Control Compartment
21. Lower Control Compartment
22. Fuel Tank
23. Fuel Filler
24. Toilet
25. Engine Control Panel
26. Battery Switch
27. Control Console
28. Air Brake Valve
29. Electric Cab Heater
30. Side Strip Heaters
31. Sliding Seats
32. Hand Brake
33. Equipment Air Filters
34. Air Duct
35. Eddy Current Clutch

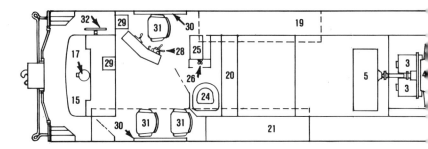

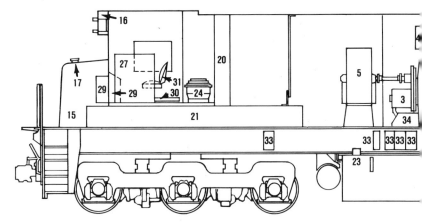

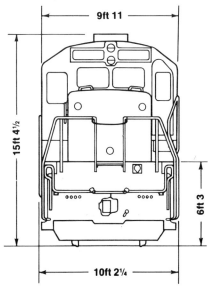

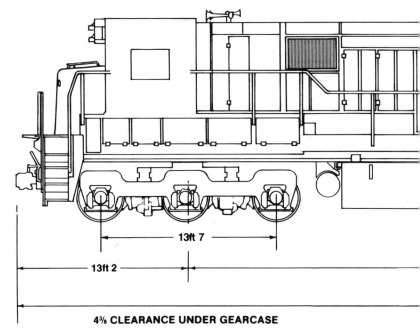

4⅜ CLEARANCE UNDER GEARCASE

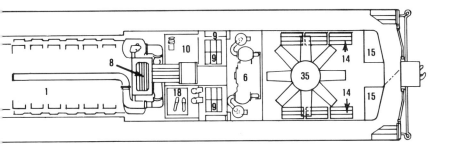
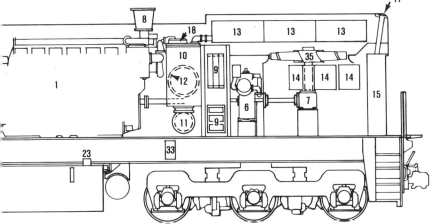
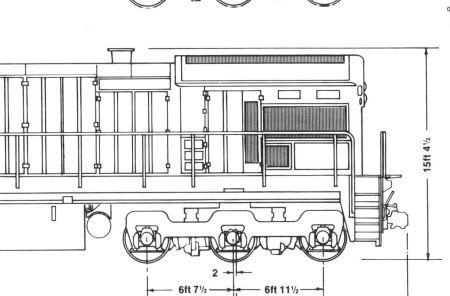
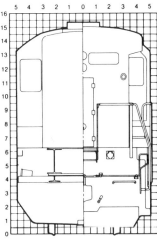

Maximum Equipment Diagram

Right half-section through exhaust stack.
Left half-section facing rear end of locomotive.
Maximum tolerance on height ± 1½ inches.

TRACTION MOTORS — GE-752AF traction motors are furnished. They are direct current, series wound, and separately ventilated by the clean air system. The armatures are mounted in anti-friction bearings. Motors drive through single-reduction spur gearing. They are supported by the axles to which they are geared and by resilient nose suspensions on truck transoms.

TRACTION ALTERNATOR — One General Electric GTA-11 traction alternator is mounted directly on the engine. It is an alternating current, single anti-friction bearing, separately-excited machine. The output is rectified by a full wave rectifier.

CONTROL — Railway type single-end control is provided. Control devices are grouped in two steel compartments, fitted with access doors. Reverser, braking switch and line contactors are electro-pneumatically operated. Other contactors are magnetically operated. Circuit breaker-type switches are used in control circuits where overcurrent protection is required. Sanding is train-lined automatically. Motor transition is automatic.

EXCITER AND BATTERY-CHARGING GENERATOR — Two Model GY27 exciters are gear-driven from the traction alternator. One provides controlled excitation of the alternator field. The other furnishes power at regulated potential for battery-charging, lighting and control.

STORAGE BATTERY — A 32-cell lead acid type, 420 ampere hour battery starts the engine and furnishes power for lights and other auxiliaries when the engine is shut down.

WHEELSLIP CORRECTION — The SENTRY Adhesion Wheelslip System uses comparative motor speed as a means of detecting slips. Highly sensitive inductive speed sensors are indexed on a special gear mounted on the motor shaft. Slip is corrected by automatic application of sand and reduction of power. Compensation for wheel size variations is automatic.

GROUND RELAY PROTECTION — If a ground occurs, engine speed returns to idle, power is removed, the alarm bell rings, and visual indication is given to the operator.

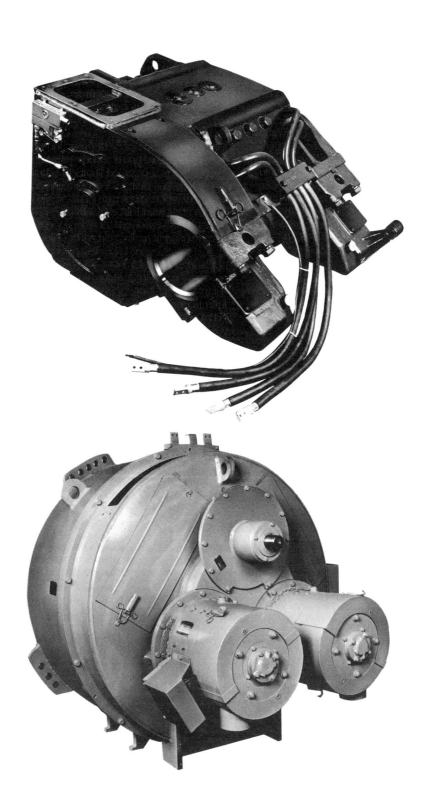

performance characteristics

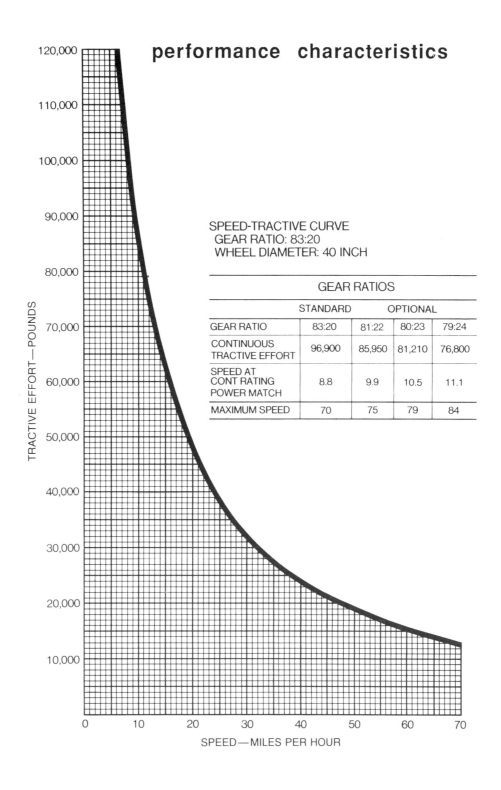

SPEED-TRACTIVE CURVE
GEAR RATIO: 83:20
WHEEL DIAMETER: 40 INCH

	GEAR RATIOS			
	STANDARD	OPTIONAL		
GEAR RATIO	83:20	81:22	80:23	79:24
CONTINUOUS TRACTIVE EFFORT	96,900	85,950	81,210	76,800
SPEED AT CONT RATING POWER MATCH	8.8	9.9	10.5	11.1
MAXIMUM SPEED	70	75	79	84

B36-7 Specifications

ratings, weights, dimensions

RATINGS
Continuous horsepower to generator for traction under standard conditions
with a General Electric 16-cylinder Model FDL16 engine .... 3600 hp
Continuous tractive effort with 83:20 gear ratio .... 64,600 lb.
Locomotive speed with 83:20 gear ratio, 40-inch wheels .... 70 mph
  (other gear ratios are available)

WEIGHTS
Minimum locomotive (fully loaded) .... 259,800 lb.
Per axle (fully loaded) .... 64,950 lb.
Locomotive weight subject to manufacturing tolerance of ± 2%
Modifications may increase weights.

WHEEL ARRANGEMENT .... B-B

MAJOR DIMENSIONS
Length inside knuckles .... 61 ft. 2 in.
Height .... 15 ft. 4½ in.
Width over handrails .... 10 ft. 2¼ in.
Minimum radius of curvature, locomotive alone .... 150 ft. (39°)

CAPACITIES
Fuel .... 2150 gal.
Engine lubricating oil .... 380 gal.
Cooling water .... 365 gal.
Sand .... 60 cu. ft.

performance features

New Series features incorporated in the 3600-hp, four-axle B36-7 are representative of General Electric's continuing design, evaluation, and manufacturing program to improve locomotive reliability, availability, and maintainability, and to reduce total operating costs.

The high performance capability of this model addresses the needs of the railroad industry for ever-increasing productivity.

DESIGN—The locomotive operating cab provides optimum visibility for operating in either direction. The control station is at the right side with the short hood leading.

POWER—A four-stroke-cycle, turbocharged diesel engine is the power source. An alternator, directly connected to the engine, furnishes power to axle-mounted traction motors. Full utilization of the engine's horsepower is available throughout the speed range of the locomotive.

OPERATION—A master controller as well as independent and automatic air brake valves are conveniently located to permit operation with either end leading. Direction of motion is controlled by a reverse lever. Throttle and reverse levers are interlocked to prevent operation of the reverser unless the throttle handle is in the off position.

SAFETY—Design of the locomotive reflects General Electric's continuing concern for safety. All safety appliances are in accordance with General Electric's interpretation of current FRA regulations.

TESTING—All General Electric locomotives are manufactured to stringent quality standards. Component parts are given standard commercial tests before assembly on the locomotive. Control wiring is checked by observing the sequence of contactor and relay operation and by testing for continuity of circuit between terminals. High-potential tests of traction and control circuits are made in accordance with current U.S.A. standards. Air brake tests assure proper operation. The power plant is tested at full load to check alternator and engine, including power and speed.

PAINTING—Durable finishes include interior gray enamel, black underframe and running gear, and special acid-resisting interior battery compartment paint. Exterior color and design are specified by the customer.

running gear

The running gear of the locomotive consists of two lateral motion swivel trucks. Center plate load is distributed by the cast-steel "floating bolster" to four rubber mounts which rest on the truck frame and provide controlled lateral motion. The cast-steel frame is supported by alloy steel coil springs over the journal boxes. Friction-type snubbers damp vertical oscillation.

WHEELS—Multiple-wear, rolled steel wheels with AAR tread and flange contour are furnished.

AXLES—Forged steel axles, conforming to AAR material specifications are provided.

JOURNALS—Journals are equipped with sealed grease-lubricated roller bearings. Pedestal openings on the truck frame have renewable non-metallic wear plates.

CENTER PLATES—Equipped with non-metallic liners, and protected by dust guards.

SAFETY HOOKS—Minimize slewing in case of derailment and permit trucks to be lifted with the locomotive superstructure.

locomotive brakes

AIR BRAKES—Schedule 26L air brake equipment with 26F control valve is furnished. The air brakes may be operated independently or with train brakes. Connections for furnishing compressed air to the train brakes are provided at each end of the locomotive.

BREAK-IN-TWO PROTECTION—Prevents the possible release of brakes from an emergency application initiated in the train with the brake valve handle in its release position.

COMPRESSOR—One 3-cylinder, 2-stage water-cooled engine-driven air compressor furnishes air for the locomotive and train braking systems.

RESERVOIRS—Reservoir capacity of 56,000 cu. in. is furnished for storing and cooling air for the brake system. Both main reservoirs are equipped with automatic drain valves.

BRAKE EQUIPMENT—Brake cylinders are supported by the truck frames and operate fully equalized brake rigging. Rigging is furnished with hardened steel bushings and adjustment to compensate for wheel and shoe wear.

HAND BRAKE—Is located on the outside of the short hood and provided to hold the locomotive at standstill.

DYNAMIC BRAKING—Brakes the locomotive electrically using the traction motors as generators and dissipating the electric power in resistors. Interlock is included to prevent application of air brakes on a locomotive in dynamic braking when automatic air is applied to the train. With standard GE braking, engine speed is automatically matched to the cooling requirements as dictated by braking level.

underframe

The welded underframe is made of rolled steel sections and plate. Hoods, cabs, equipment, and tanks are supported by the main frame members. Space between these members is enclosed to form an air duct which distributes clean air throughout the locomotive.

WEARPLATES—Renewable, wear-resistant steel plates are applied to side bearing pads, and draft gear housing.

COUPLERS—AAR type E top-operated couplers with NC-391 rubber-cushioned draft gear and alignment control are provided at each locomotive end.

PILOT AND SIDE STEPS—A pilot is at each end of the locomotive. Side steps provide access to the platform.

LIFTING AND JACKING—Four jacking pads in combination with lugs for cable slings are provided on the side bolsters.

FUEL TANK—A heavy gage welded steel tank is bolted to the underframe between the trucks. The tank is provided with baffle plates, clean-out plug, water drains, and vent. Filler connections and fuel level gages are furnished on each side of the locomotive. Emergency fuel shutoffs are provided.

superstructure

The welded steel superstructure consists of a short front hood, operator's cab, engine hood, and radiator compartment. Engine hood is bolted to the underframe and is removable.

SHORT HOOD—The short hood contains a top-serviced sandbox. A door in the front bulkhead of the cab provides access. Classification lights are mounted on this hood. A ventilator is provided.

OPERATOR'S CAB—Sides and roof are insulated and steel-lined. The floor, raised above the underframe, is covered with high density hardboard. The cab has windows in the front and rear. Two-pane center windows on each side have sliding sash equipped with latches. Doors in diagonally opposite corners of the cab provide access to walkways on both sides of the locomotive. They have windows, weather stripping, and locks. All cab glazing is certified in compliance with FRA safety glazing standards. Headlights and number boxes are arranged on the outside above the front windows. Electric cab heat is provided.

WALKWAYS—Walkways with handrails and non-skid treads are at each end of the locomotive and along the hoods.

ENGINE COMPARTMENT—Encloses the engine, and traction alternator. Complete access to this equipment is provided by doors the full height and length of the hood extending along both sides of the locomotive. Doors in the roof provide overhead access to cylinders. Detachable roof sections permit removal of equipment.

RADIATOR COMPARTMENT—Contains the radiators, fan and gearbox, compressor, blower and engine air cleaners. The compressor is enclosed in the radiator compartment with free air access from the engine compartment. The radiators are roof-mounted. A reinforced screen over the air outlet opening is removable to permit removal of the radiators, fan, and gearbox. Dynamic braking grids are mounted along each side of the radiator compartment. An end section holds a sandbox, serviced from the roof. Rear headlights, classification lights, and number boxes are mounted on this section.

EQUIPMENT COMPARTMENTS—Main propulsion control equipment is located on the left side of the locomotive beneath the operator's cab. This compartment, maintained under positive air pressure to keep out dirt and water, contains contactors, reverser, and braking switch. Excitation and other panels and devices are located in a compartment behind the operator's cab. The compartment is gasketed to prevent entrance of dust. It can be accessed from either walkway but is not accessible from the operator's cab. Air brake devices are located in a compartment along the right side of the locomotive under the operator's cab. Battery trays are located in a box with hinged top doors for ease of servicing.

VENTILATION—Filtered air is provided through self-cleaning air cleaners located in the underframe. Clean air is delivered under pressure for equipment cooling and pressurization, and cab ventilation. Engine air is cleaned by self-cleaning air cleaners and by General Electric paper filters.

power plant

DIESEL ENGINE—Type—one General Electric FDL; cylinder arrangement—45°V; stroke cycle—4; bore and stroke—9 X 10½ inches; RPM—1050; turbocharger—one.

GOVERNOR—Self-contained, electro-hydraulic Woodward PG type governor automatically regulates horsepower output at each throttle setting.

OVERSPEED PROTECTION—The engine automatically shuts down if speed exceeds maximum rated rpm by 10%.

COOLING SYSTEM—A gear-driven centrifugal pump integral with the diesel engine circulates cooling water through the engine, turbocharger, air intercoolers, self-draining radiators, lubricating oil cooler and air compressor. The tank is equipped with a sight gage to indicate water level and with screens which provide maximum filtration of debris and scale. The system is pressurized. Abnormally low water pressure automatically shuts down the engine.

TEMPERATURE CONTROL—A solid-state fluid amplifier control system and variable speed radiator fan automatically maintain cooling system temperature. The fluid amplifier regulates the flow of cooling water through the radiator sections. The radiator fan eddy current clutch matches fan speed/horsepower to cooling requirements.

FUEL SYSTEM—A motor-driven pump transfers fuel from the tank through a strainer and filter to the injection pumps. Each cylinder is equipped with a high pressure fuel injection pump and injector.

LUBRICATING SYSTEM—A single pressure-regulated system is supplied by a gear type pump integral with the diesel engine. A lubricating oil reservoir is located in the engine subbase. Suction strainer, lubricating oil filters, and water-cooled oil cooler are provided. Abnormally low lubricating oil pressure or abnormally high crankcase pressure automatically shuts down the engine.

ENGINE STARTING—The engine is cranked by the two GY27's from storage battery power.

HORSEPOWER OUTPUT—Horsepower input to the alternator for traction is provided under AAR standard conditions with specified fuel and lubricating oil.

LOW IDLE—Has been provided to reduce engine speed during prolonged periods of idle operation.

location of major systems

1. Engine - GE Model 7FDL16
2. Alternator
3. Auxiliary Generator
4. Rectifiers
5. Equipment Blower
6. Air Compressor
7. Radiator Fan Gear Unit
8. Engine Exhaust Stack
9. Engine Air Filters
10. Engine Water Tank
11. Lube Oil Cooler
12. Lube Oil Filter
13. Radiator
14. Braking Resistors
15. Sand Box
16. Number Box
17. Sand Fill
18. Fluid Amplifier
19. Battery Box
20. Upper Control Compartment
21. Lower Control Compartment
22. Fuel Tank
23. Fuel Filler
24. Toilet
25. Engine Control Panel
26. Battery Switch
27. Control Console
28. Air Brake Valve
29. Electric Cab Heater
30. Side Strip Heaters
31. Sliding Seats
32. Hand Brake
33. Equipment Air Filters
34. Air Duct
35. Eddy Current Clutch

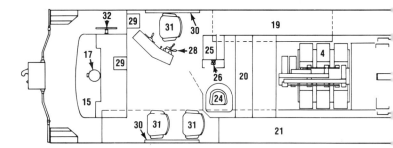

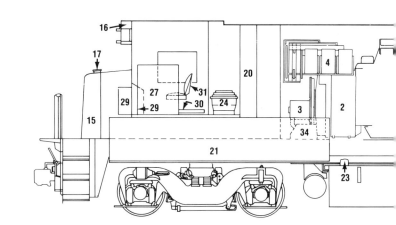

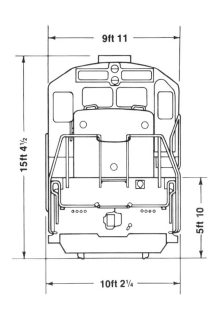

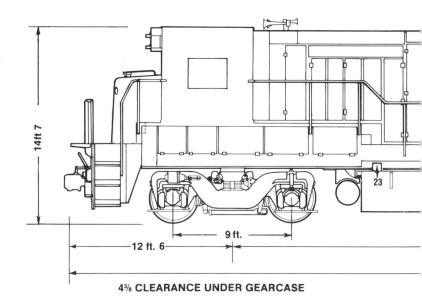

4³⁄₈ CLEARANCE UNDER GEARCASE

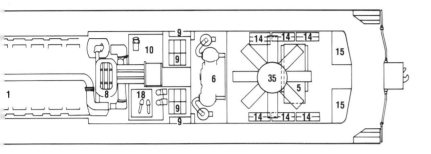

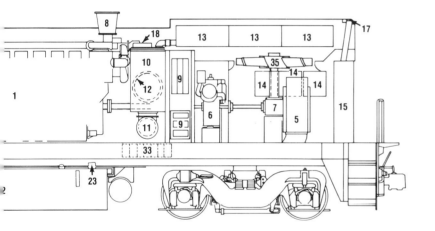

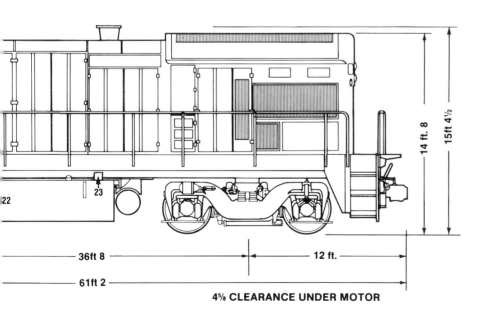

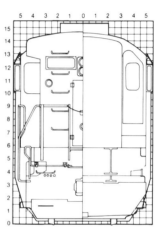

Maximum Equipment Diagram

Right half-section through exhaust stack.
Left half-section facing rear end of locomotive.
Maximum tolerance on height: ± 1½ inches.

electric transmission

TRACTION MOTORS—GE-752AF traction motors are furnished. They are direct current, series wound, and separately ventilated by the clean air system. The armatures are mounted in anti-friction bearings. Motors drive through single-reduction spur gearing. They are supported by the axles to which they are geared and by resilient nose suspensions on truck transoms.

TRACTION ALTERNATOR—One General Electric GTA-24 traction alternator is mounted directly on the engine. It is an alternating current, single anti-friction bearing, separately-excited machine. The output is rectified by a full wave rectifier.

CONTROL—Railway type single-end control is provided. Control devices are grouped in two steel compartments, fitted with access doors. Reverser, braking switch and line contactors are electro-pneumatically operated. Other contactors are magnetically operated. Circuit breaker-type switches are used in control circuits where overcurrent protection is required. Sanding is train-lined automatically. Alternator transition is automatic.

EXCITER AND BATTERY-CHARGING GENERATOR—Two Model GY27 exciters are gear-driven from the traction alternator. One provides controlled excitation of the alternator field. The other furnishes power at regulated potential for battery-charging, lighting and control.

STORAGE BATTERY—A 32-cell lead acid type, 420 ampere hour battery starts the engine and furnishes power for lights and other auxiliaries when the engine is shut down.

WHEELSLIP CORRECTION—The SENTRY Adhesion Wheelslip System uses comparative motor speed as a means of detecting slips. Highly sensitive inductive speed sensors are indexed on a special gear mounted on the motor shaft. Slip is corrected by automatic application of sand and reduction of power. Compensation for wheel size variations is automatic.

GROUND RELAY PROTECTION—If a ground occurs, engine speed returns to idle, power is removed, the alarm bell rings, and visual indication is given to the operator.

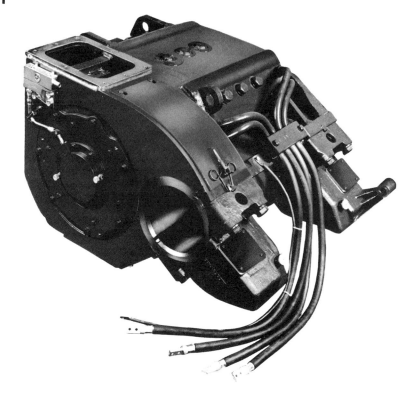

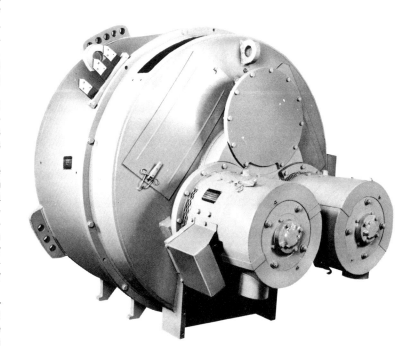

operating controls

Controls and instruments are grouped at the operator's station and auxiliary panels in the operator's cab. The controller/console meets all AAR requirements.

OPERATING CONTROLS
Controller with throttle, dynamic braking
  handle and reverser-selector levers
Engine start push button
Engine stop push button and emergency fuel shutoff
Brake valves
Bell ringer valve
Air horn valve
Window wiper valves
Circuit breakers and switches
Emergency multiple unit engine stop switch
Sander switch
Lead axle sander switch
Ground relay reset push button
Engine control switch

INSTRUMENTS
Two brake gages with test fittings
Dual reading load meter
Mechanical-type speed recorder and odometer
Fuel oil pressure gage.
WARNING INDICATORS
Warning light—wheelslip, PCS,
  annunciator light
Alarm bell and warning light
  High crankcase pressure
  Ground relay
  Engine overtemperature
  No battery charge
  Governor shutdown
  Engine air filters
  Hot diodes

locomotive accessories

BELL—Cast iron with air-operated ringer and operating valve.
CAB HEAT—Electric.
CLASSIFICATION LIGHTS—Two, 3-aspect electric lights at each end of the locomotive.
CLOTHES HOOKS—Provided in operator's cab.
CONDUCTOR'S EMERGENCY VALVE—Standard AAR location.
EMERGENCY FUEL SHUTOFF—Three pushbuttons, one on each side of the underframe and one in the operator's cab.
FIRE EXTINGUISHERS—Two, 20-pound dry chemical, one at each end of the locomotive.
FUEL GAGES—One on each side and one on each end for upper and lower level readings.
HEADLIGHTS—At each end of the locomotive to comply with FRA requirements. Dimming control provided with the lights.
HORN—One, 3-tone with two bells forward and one to the rear.
INTERIOR LIGHTS—For illuminating the operator's cab, nose cab, and instruments.
SANDERS—Eight, pneumatically operated, to sand ahead of the lead wheels of each truck in each direction.
SEATS—Two, wall-mounted, adjustable for height and for operating in either direction. Cushioned arm rests provided at side windows.
STEP LIGHTS—Four, one for each side step.
SUN VISOR—Adjustable-type.
WATER TEMPERATURE GAGE—Located in the engine compartment.
WINDOW WIPERS—Six, air-operated, on front and rear windows of operator's cab.

modifications

The following modifications may increase locomotive weight, dimensions and price.

ADDITIONAL FUEL—1100 gallons for a maximum fuel capacity of 3250 gallons.
AUTOMATIC SANDING—Sanding in either direction initiated in the event of emergency brake application.
AWNINGS—Over window on each side of cab.
BALLAST—Locomotive can be ballasted up to 280,000 pounds.
BATTERY CHARGING AMMETER—Mounted on engine control panel.
BATTERY CHARGING RECEPTACLE—One 150-ampere type.
BRAKE SHOES—To meet customer requirements.
CAB SIGNAL EQUIPMENT—Train control cab signal or train speed control equipment as now used on various railroads.
COUPLER—AAR Type F in place of Type E. NC-390 draft gear required.
DELUXE CAB SEATS—Upholstered with armrests.
DRAFT GEAR—NC-390 equipment furnished.
DYNAMIC BRAKING—Furnished as standard equipment.
EXTENDED RANGE DYNAMIC BRAKE—Equipment provided with dynamic braking to obtain greater braking effort at low train speed.
EXTRA CAB SEAT—Third seat in operator's cab.
FIRE EXTINGUISHER—To meet customer requirements.
FUEL LEVEL GAGES—Dial-type gages provided on both sides of the tank near the filler openings.
GEAR RATIOS—Optional gear ratios available as follows:

Gears	81:22	80:23	79:24
Ratio	3.68	3.48	3.29
Max. mph	75	79	84

LONG HOOD LEADING CONTROL—Control station mounted for operation with the long hood leading.
MULTIPLE UNIT CONTROL—M.U. control furnished with M.U. crosswalks to operate two or more units with 26L or 24L air brakes from one cab.
ON-BOARD LOAD TESTING—To permit loading for test purposes on the dynamic braking grids.
OVERSPEED PROTECTION—To return engine to idle, cut off power, and make automatic brake application.
RETENTION TANK—To hold liquids collected by platform drains.
SAFETY CONTROL—Safety deadman control including foot pedal valve, time delay, warning whistle, and brake application.
TOILET—Electric incinerating, biodegradable, or retention type. Located in compartment at the rear of the operator's cab.
TOOL BOX
TRAIN COMMUNICATION—As used by various railroads.
TRUCK JOURNAL BEARINGS—Hyatt bearings in place of the standard Timken GG bearings.
TWO-STATION CONTROL—For operating the locomotive from either of two diagonally opposite positions in the operator's cab.
VISUAL WARNING SIGNAL LIGHTS—Located at each end of the locomotive.
WATER COOLER/REFRIGERATOR—Floor-mounted in operator's cab.
WINDSHIELD WINGS—Wind deflectors; one in front and rear of each side window.

performance characteristics

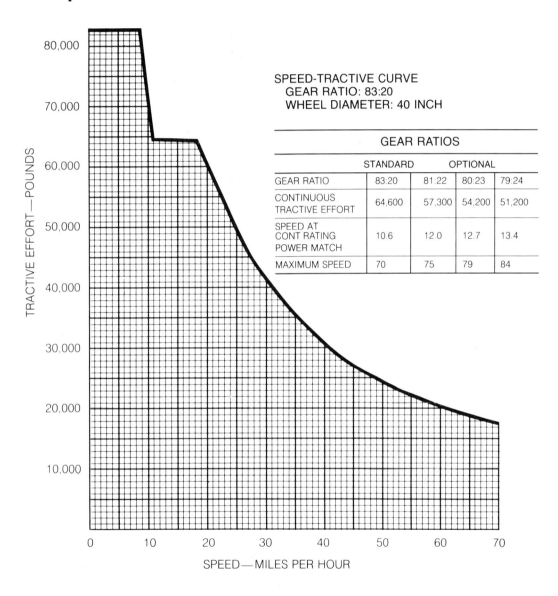

SPEED-TRACTIVE CURVE
GEAR RATIO: 83:20
WHEEL DIAMETER: 40 INCH

	GEAR RATIOS			
	STANDARD	OPTIONAL		
GEAR RATIO	83:20	81:22	80:23	79:24
CONTINUOUS TRACTIVE EFFORT	64,600	57,300	54,200	51,200
SPEED AT CONT RATING POWER MATCH	10.6	12.0	12.7	13.4
MAXIMUM SPEED	70	75	79	84

B30-7 Specifications

ratings, weights, dimensions

RATINGS
Continuous horsepower to generator for traction under standard conditions
with a General Electric 16-cylinder Model FDL16 engine .... 3000 hp
Continuous tractive effort with 83:20 gear ratio .... 61,000 lb.
Locomotive speed with 83:20 gear ratio, 40-inch wheels .... 70 mph
   (other gear ratios are available)

WEIGHTS
Minimum locomotive (fully loaded) .... 259,000 lb.
Per axle (fully loaded) .... 64,750 lb.
Locomotive weight subject to manufacturing tolerance of ± 2%
Modifications may increase weights.

WHEEL ARRANGEMENT .... B-B

MAJOR DIMENSIONS
Length inside knuckles .... 61 ft. 2 in.
Height .... 15 ft. 4½ in.
Width over handrails .... 10 ft. 2¼ in.
Minimum radius of curvature, locomotive alone .... 150 ft. (39°)

CAPACITIES
Fuel .... 2150 gal.
Engine lubricating oil .... 380 gal.
Cooling water .... 365 gal.
Sand .... 60 cu. ft.

performance features

New Series features incorporated in the 3000-hp, four-axle B30-7 are representative of General Electric's continuing design, evaluation, and manufacturing program to improve locomotive reliability, availability, and maintainability, and to reduce total operating costs.

The high performance capability of this model addresses the needs of the railroad industry for ever-increasing productivity.

DESIGN—The locomotive operating cab provides optimum visibility for operating in either direction. The control station is at the right side with the short hood leading.

POWER—A four-stroke-cycle, turbocharged diesel engine is the power source. An alternator, directly connected to the engine, furnishes power to axle-mounted traction motors. Full utilization of the engine's horsepower is available throughout the speed range of the locomotive.

OPERATION—A master controller as well as independent and automatic air brake valves are conveniently located to permit operation with either end leading. Direction of motion is controlled by a reverse lever. Throttle and reverse levers are interlocked to prevent operation of the reverser unless the throttle handle is in the off position.

SAFETY—Design of the locomotive reflects General Electric's continuing concern for safety. All safety appliances are in accordance with General Electric's interpretation of current FRA regulations.

TESTING—All General Electric locomotives are manufactured to stringent quality standards. Component parts are given standard commercial tests before assembly on the locomotive. Control wiring is checked by observing the sequence of contactor and relay operation and by testing for continuity of circuit between terminals. High-potential tests of traction and control circuits are made in accordance with current U.S.A. standards. Air brake tests assure proper operation. The power plant is tested at full load to check alternator and engine, including power and speed.

PAINTING—Durable finishes include interior gray enamel, black underframe and running gear, and special acid-resisting interior battery compartment paint. Exterior color and design are specified by the customer.

running gear

The running gear of the locomotive consists of two lateral motion swivel trucks. Center plate load is distributed by the cast-steel "floating bolster" to four rubber mounts which rest on the truck frame and provide controlled lateral motion. The cast-steel frame is supported by alloy steel coil springs over the journal boxes. Friction-type snubbers damp vertical oscillation.

WHEELS—Multiple-wear, rolled steel wheels with AAR tread and flange contour are furnished.

AXLES—Forged steel axles, conforming to AAR material specifications are provided.

JOURNALS—Journals are equipped with sealed grease-lubricated roller bearings. Pedestal openings on the truck frame have renewable non-metallic wear plates.

CENTER PLATES—Equipped with non-metallic liners, and protected by dust guards.

SAFETY HOOKS—Minimize slewing in case of derailment and permit trucks to be lifted with the locomotive superstructure.

location of major systems

1. Engine - GE Model 7FDL16
2. Alternator
3. Auxiliary Generator
4. Rectifiers
5. Equipment Blower
6. Air Compressor
7. Radiator Fan Gear Unit
8. Engine Exhaust Stack
9. Engine Air Filters
10. Engine Water Tank
11. Lube Oil Cooler
12. Lube Oil Filter
13. Radiator
14. Braking Resistors
15. Sand Box
16. Number Box
17. Sand Fill
18. Fluid Amplifier
19. Battery Box
20. Upper Control Compartment
21. Lower Control Compartment
22. Fuel Tank
23. Fuel Filler
24. Toilet
25. Engine Control Panel
26. Battery Switch
27. Control Console
28. Air Brake Valve
29. Electric Cab Heater
30. Side Strip Heaters
31. Sliding Seats
32. Hand Brake
33. Equipment Air Filters
34. Air Duct
35. Eddy Current Clutch

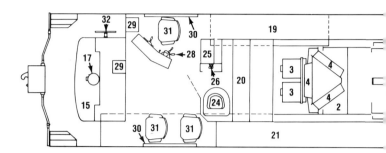

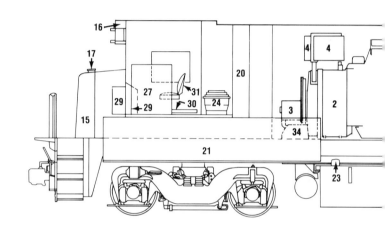

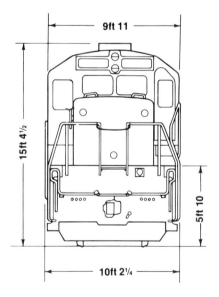

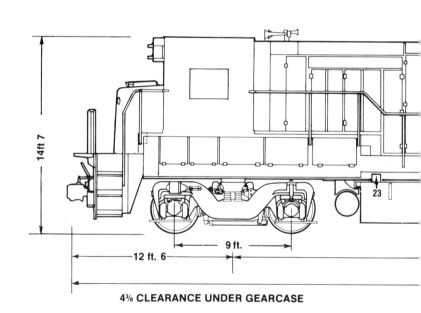

4 3/8 CLEARANCE UNDER GEARCASE

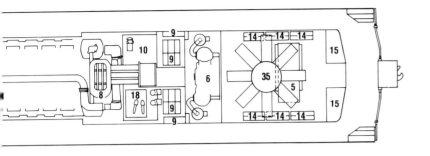

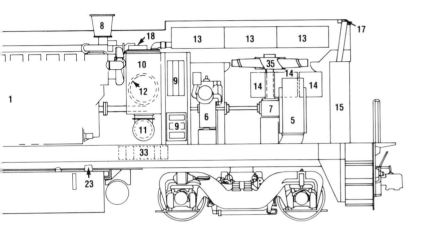

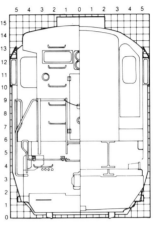

Maximum Equipment Diagram
Right half-section through exhaust stack.
Left half-section facing rear end of locomotive.
Maximum tolerance on height: ±1½ inches.

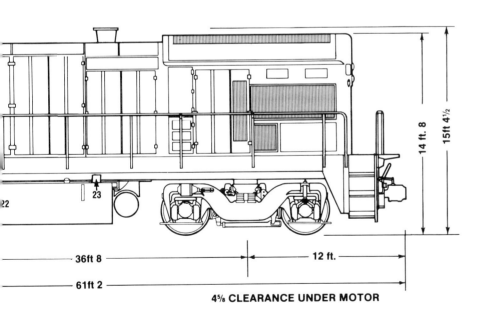

-119-

power plant

DIESEL ENGINE—Type—one General Electric FDL; cylinder arrangement—45°V; stroke cycle—4; bore and stroke—9 X 10½ inches; RPM—1050; turbocharger—one.

GOVERNOR—Self-contained, electro-hydraulic Woodward PG type governor automatically regulates horsepower output at each throttle setting.

OVERSPEED PROTECTION—The engine automatically shuts down if speed exceeds maximum rated rpm by 10%.

COOLING SYSTEM—A gear-driven centrifugal pump integral with the diesel engine circulates cooling water through the engine, turbocharger, air intercoolers, self-draining radiators, lubricating oil cooler and air compressor. The tank is equipped with a sight gage to indicate water level and with screens which provide maximum filtration of debris and scale. The system is pressurized. Abnormally low water pressure automatically shuts down the engine.

TEMPERATURE CONTROL—A solid-state fluid amplifier control system, thermostatically operated, automatically maintains temperature by regulating the flow of cooling water through the radiator sections.

FUEL SYSTEM—A motor-driven pump transfers fuel from the tank through a strainer and filter to the injection pumps. Each cylinder is equipped with a high pressure fuel injection pump and injector.

LUBRICATING SYSTEM—A single pressure-regulated system is supplied by a gear type pump integral with the diesel engine. A lubricating oil reservoir is located in the engine sub-base. Suction strainer, lubricating oil filters, and water-cooled oil cooler are provided. Abnormally low lubricating oil pressure or abnormally high crankcase pressure automatically shuts down the engine.

ENGINE STARTING—The engine is cranked by the two GY27's from storage battery power.

HORSEPOWER OUTPUT—Horsepower input to the alternator for traction is provided under AAR standard conditions with specified fuel and lubricating oil.

LOW IDLE—Has been provided to reduce engine speed during prolonged periods of idle operation.

electric transmission

TRACTION MOTORS—GE-752AF traction motors are furnished. They are direct current, series wound, and separately ventilated by the clean air system. The armatures are mounted in anti-friction bearings. Motors drive through single-reduction spur gearing. They are supported by the axles to which they are geared and by resilient nose suspensions on truck transoms.

TRACTION ALTERNATOR—One General Electric GTA-11 traction alternator is mounted directly on the engine. It is an alternating current, single anti-friction bearing, separately-excited machine. The output is rectified by a full wave rectifier.

CONTROL—Railway type single-end control is provided. Control devices are grouped in two steel compartments, fitted with access doors. Reverser, braking switch and line contactors are electro-pneumatically operated. Other contactors are magnetically operated. Circuit breaker-type switches are used in control circuits where overcurrent protection is required. Sanding is train-lined automatically. Transition is automatic.

EXCITER AND BATTERY-CHARGING GENERATOR—Two Model GY27 exciters are gear-driven from the traction alternator. One provides controlled excitation of the alternator field. The other furnishes power at regulated potential for battery-charging, lighting and control.

STORAGE BATTERY—A 32-cell lead acid type, 420 ampere hour battery starts the engine and furnishes power for lights and other auxiliaries when the engine is shut down.

WHEELSLIP CORRECTION—Wheelslip is automatically detected by comparison of output signals from current measuring reactors located in each motor circuit. Slip is automatically corrected by automatic application of sand and reduction of power.

GROUND RELAY PROTECTION—If a ground occurs, engine speed returns to idle, power is removed, the alarm bell rings, and visual indication is given to the operator.

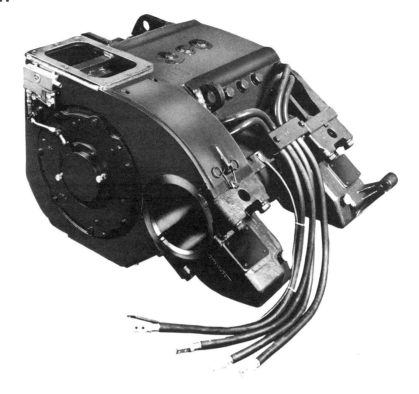

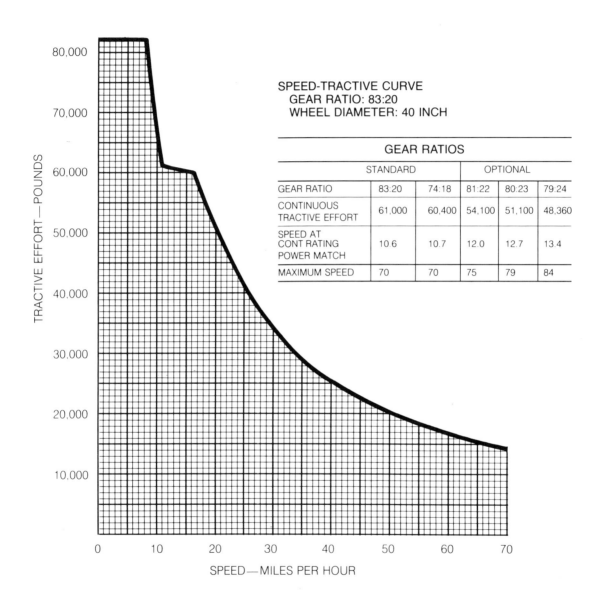

B23-7 Specifications

ratings, weights, dimensions

RATINGS
Continuous horsepower to generator for traction under standard conditions
with a General Electric 12-cylinder Model FDL12 engine .... 2250 hp
Continuous tractive effort with 83:20 gear ratio .... 61,000 lb.
Locomotive speed with 83:20 gear ratio, 40-inch wheels .... 70 mph
  (other gear ratios are available)

WEIGHTS
Minimum locomotive (fully loaded) .... 253,000 lb.
Per axle (fully loaded) .... 63,250 lb.
Locomotive weight subject to manufacturing tolerance of ± 2%
Modifications may increase weights.

WHEEL ARRANGEMENT .... B-B

MAJOR DIMENSIONS
Length inside knuckles .... 61 ft. 2 in.
Height .... 15 ft. 4½ in.
Width over handrails .... 10 ft. 2¼ in.
Minimum radius of curvature, locomotive alone .... 150 ft. (39°)

CAPACITIES
Fuel .... 2150 gal.
Engine lubricating oil .... 350 gal.
Cooling water .... 300 gal.
Sand .... 60 cu. ft.

performance features

New Series features incorporated in the 2300-hp, four-axle B23-7 are representative of General Electric's continuing design, evaluation, and manufacturing program to improve locomotive reliability, availability, and maintainability, and to reduce total operating costs.

The high performance capability of this model addresses the needs of the railroad industry for ever-increasing productivity.

DESIGN—The locomotive operating cab provides optimum visibility for operating in either direction. The control station is at the right side with the short hood leading.

POWER—A four-stroke-cycle, turbocharged diesel engine is the power source. An alternator, directly connected to the engine, furnishes power to axle-mounted traction motors. Full utilization of the engine's horsepower is available throughout the speed range of the locomotive.

OPERATION—A master controller as well as independent and automatic air brake valves are conveniently located to permit operation with either end leading. Direction of motion is controlled by a reverse lever. Throttle and reverse levers are interlocked to prevent operation of the reverser unless the throttle handle is in the off position.

SAFETY—Design of the locomotive reflects General Electric's continuing concern for safety. All safety appliances are in accordance with General Electric's interpretation of current FRA regulations.

TESTING—All General Electric locomotives are manufactured to stringent quality standards. Component parts are given standard commercial tests before assembly on the locomotive. Control wiring is checked by observing the sequence of contactor and relay operation and by testing for continuity of circuit between terminals. High-potential tests of traction and control circuits are made in accordance with current U.S.A. standards. Air brake tests assure proper operation. The power plant is tested at full load to check alternator and engine, including power and speed.

PAINTING—Durable finishes include interior gray enamel, black underframe and running gear, and special acid-resisting interior battery compartment paint. Exterior color and design are specified by the customer.

running gear

The running gear of the locomotive consists of two lateral motion swivel trucks. Center plate load is distributed by the cast-steel "floating bolster" to four rubber mounts which rest on the truck frame and provide controlled lateral motion. The cast-steel frame is supported by alloy steel coil springs over the journal boxes. Friction-type snubbers damp vertical oscillation.

WHEELS—Multiple-wear, rolled steel wheels with AAR tread and flange contour are furnished.

AXLES—Forged steel axles, conforming to AAR material specifications are provided.

JOURNALS—Journals are equipped with sealed grease-lubricated roller bearings. Pedestal openings on the truck frame have renewable non-metallic wear plates.

CENTER PLATES—Equipped with non-metallic liners, and protected by dust guards.

SAFETY HOOKS—Minimize slewing in case of derailment and permit trucks to be lifted with the locomotive superstructure.

locomotive brakes

AIR BRAKES—Schedule 26L air brake equipment with 26F control valve is furnished. The air brakes may be operated independently or with train brakes. Connections for furnishing compressed air to the train brakes are provided at each end of the locomotive.

BREAK-IN-TWO PROTECTION—Prevents the possible release of brakes from an emergency application initiated in the train with the brake valve handle in its release position.

COMPRESSOR—One 3-cylinder, 2-stage water-cooled engine-driven air compressor furnishes air for the locomotive and train braking systems.

RESERVOIRS—Reservoir capacity of 56,000 cu. in. is furnished for storing and cooling air for the brake system. Both main reservoirs are equipped with automatic drain valves.

BRAKE EQUIPMENT—Brake cylinders are supported by the truck frames and operate fully equalized brake rigging. Rigging is furnished with hardened steel bushings and adjustment to compensate for wheel and shoe wear.

HAND BRAKE—Is located on the outside of the short hood and provided to hold the locomotive at standstill.

DYNAMIC BRAKING—Brakes the locomotive electrically using the traction motors as generators and dissipating the electric power in resistors. Interlock is included to prevent application of air brakes on a locomotive in dynamic braking when automatic air is applied to the train. With standard GE braking, engine speed is automatically matched to the cooling requirements as dictated by braking level.

underframe

The welded underframe is made of rolled steel sections and plate. Hoods, cabs, equipment, and tanks are supported by the main frame members. Space between these members is enclosed to form an air duct which distributes clean air throughout the locomotive.

WEARPLATES—Renewable, wear-resistant steel plates are applied to side bearing pads, and draft gear housing.

COUPLERS—AAR type E top-operated couplers with NC-391 rubber-cushioned draft gear and alignment control are provided at each locomotive end.

PILOT AND SIDE STEPS—A pilot is at each end of the locomotive. Side steps provide access to the platform.

LIFTING AND JACKING—Four jacking pads in combination with lugs for cable slings are provided on the side bolsters.

FUEL TANK—A heavy gage welded steel tank is bolted to the underframe between the trucks. The tank is provided with baffle plates, clean-out plug, water drains, and vent. Filler connections and fuel level gages are furnished on each side of the locomotive. Emergency fuel shutoffs are provided.

superstructure

The welded steel superstructure consists of a short front hood, operator's cab, engine hood, and radiator compartment. Engine hood is bolted to the underframe and is removable.

SHORT HOOD—The short hood contains a top-serviced sandbox. A door in the front bulkhead of the cab provides access. Classification lights are mounted on this hood. A ventilator is provided.

OPERATOR'S CAB—Sides and roof are insulated and steel-lined. The floor, raised above the underframe, is covered with high density hardboard. The cab has windows in the front and rear. Two-pane center windows on each side have sliding sash equipped with latches. Doors in diagonally opposite corners of the cab provide access to walkways on both sides of the locomotive. They have windows, weather stripping, and locks. All cab glazing is certified in compliance with FRA safety glazing standards. Headlights and number boxes are arranged on the outside above the front windows. Electric cab heat is provided.

WALKWAYS—Walkways with handrails and non-skid treads are at each end of the locomotive and along the hoods.

ENGINE COMPARTMENT—Encloses the engine, and traction alternator. Complete access to this equipment is provided by doors the full height and length of the hood extending along both sides of the locomotive. Doors in the roof provide overhead access to cylinders. Detachable roof sections permit removal of equipment.

RADIATOR COMPARTMENT—Contains the radiators, fan and gearbox, compressor, blower and engine air cleaners. The compressor is enclosed in the radiator compartment with free air access from the engine compartment. The radiators are roof-mounted. A reinforced screen over the air outlet opening is removable to permit removal of the radiators, fan, and gearbox. Dynamic braking grids are mounted along each side of the radiator compartment. An end section holds a sandbox, serviced from the roof. Rear headlights, classification lights, and number boxes are mounted on this section.

EQUIPMENT COMPARTMENTS—Main propulsion control equipment is located on the left side of the locomotive beneath the operator's cab. This compartment, maintained under positive air pressure to keep out dirt and water, contains contactors, reverser, and braking switch. Excitation and other panels and devices are located in a compartment behind the operator's cab. The compartment is gasketed to prevent entrance of dust. It can be accessed from either walkway but is not accessible from the operator's cab. Air brake devices are located in a compartment along the right side of the locomotive under the operator's cab. Battery trays are located in a box with hinged top doors for ease of servicing.

VENTILATION—Filtered air is provided through self-cleaning air cleaners located in the underframe. Clean air is delivered under pressure for equipment cooling and pressurization, and cab ventilation. Engine air is cleaned by self-cleaning air cleaners and by General Electric paper filters.

location of major systems

1. Engine - GE Model 7FDL12
2. Alternator
3. Auxiliary Generator
4. Rectifiers
5. Equipment Blower
6. Air Compressor
7. Radiator Fan Gear Unit
8. Engine Exhaust Stack
9. Engine Air Filters
10. Engine Water Tank
11. Lube Oil Cooler
12. Lube Oil Filter
13. Radiator
14. Braking Resistors
15. Sand Box
16. Number Box
17. Sand Fill
18. Fluid Amplifier
19. Battery Box
20. Upper Control Compartment
21. Lower Control Compartment
22. Fuel Tank
23. Fuel Filler
24. Toilet
25. Engine Control Panel
26. Battery Switch
27. Control Console
28. Air Brake Valve
29. Electric Cab Heater
30. Side Strip Heaters
31. Sliding Seats
32. Hand Brake
33. Equipment Air Filters
34. Air Duct

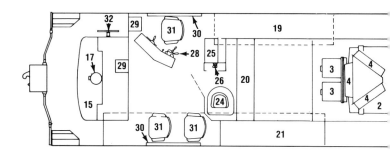

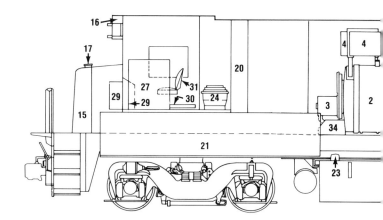

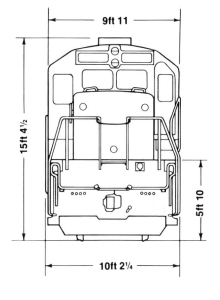

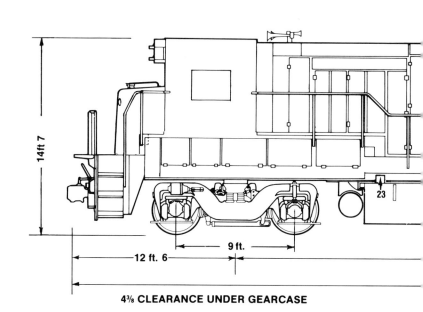

4³⁄₈ CLEARANCE UNDER GEARCASE

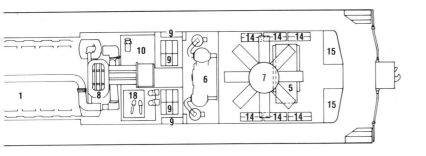

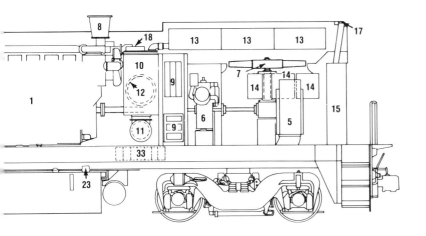

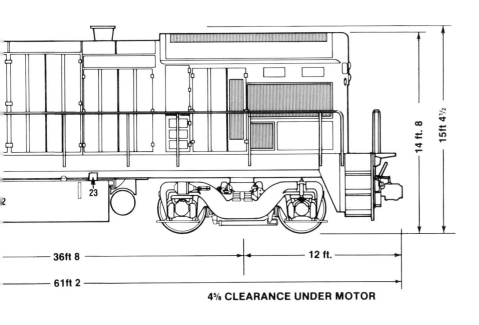

36ft 8
61ft 2

4⅝ CLEARANCE UNDER MOTOR

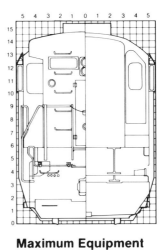

Maximum Equipment Diagram

Right half-section through exhaust stack.
Left half-section facing rear end of locomotive.
Maximum tolerance on height: ± 1½ inches.

power plant

DIESEL ENGINE—Type—one General Electric FDL; cylinder arrangement—45°V; stroke cycle—4; bore and stroke—9 X 10½ inches; RPM—1050; turbocharger—one.

GOVERNOR—Self-contained, electro-hydraulic Woodward PG type governor automatically regulates horsepower output at each throttle setting.

OVERSPEED PROTECTION—The engine automatically shuts down if speed exceeds maximum rated rpm by 10%.

COOLING SYSTEM—A gear-driven centrifugal pump integral with the diesel engine circulates cooling water through the engine, turbocharger, air intercoolers, self-draining radiators, lubricating oil cooler and air compressor. The tank is equipped with a sight gage to indicate water level and with screens which provide maximum filtration of debris and scale. The system is pressurized. Abnormally low water pressure automatically shuts down the engine.

TEMPERATURE CONTROL—A solid-state fluid amplifier control system and variable speed radiator fan automatically maintain cooling system temperature. The fluid amplifier regulates the flow of cooling water through the radiator sections. The radiator fan eddy current clutch matches fan speed/horsepower to cooling requirements.

FUEL SYSTEM—A motor-driven pump transfers fuel from the tank through a strainer and filter to the injection pumps. Each cylinder is equipped with a high pressure fuel injection pump and injector.

LUBRICATING SYSTEM—A single pressure-regulated system is supplied by a gear type pump integral with the diesel engine. A lubricating oil reservoir is located in the engine subbase. Suction strainer, lubricating oil filters, and water-cooled oil cooler are provided. Abnormally low lubricating oil pressure or abnormally high crankcase pressure automatically shuts down the engine.

ENGINE STARTING—The engine is cranked by the two GY27's from storage battery power.

HORSEPOWER OUTPUT—Horsepower input to the alternator for traction is provided under AAR standard conditions with specified fuel and lubricating oil.

LOW IDLE—Has been provided to reduce engine speed during prolonged periods of idle operation.

electric transmission

TRACTION MOTORS—GE-752AF traction motors are furnished. They are direct current, series wound, and separately ventilated by the clean air system. The armatures are mounted in anti-friction bearings. Motors drive through single-reduction spur gearing. They are supported by the axles to which they are geared and by resilient nose suspensions on truck transoms.

TRACTION ALTERNATOR—One General Electric GTA-11 traction alternator is mounted directly on the engine. It is an alternating current, single anti-friction bearing, separately-excited machine. The output is rectified by a full wave rectifier.

CONTROL—Railway type single-end control is provided. Control devices are grouped in two steel compartments, fitted with access doors. Reverser, braking switch and line contactors are electro-pneumatically operated. Other contactors are magnetically operated. Circuit breaker-type switches are used in control circuits where overcurrent protection is required. Sanding is train-lined automatically. Transition is automatic.

EXCITER AND BATTERY-CHARGING GENERATOR—Two Model GY27 exciters are gear-driven from the traction alternator. One provides controlled excitation of the alternator field. The other furnishes power at regulated potential for battery-charging, lighting and control.

STORAGE BATTERY—A 32-cell lead acid type, 420 ampere hour battery starts the engine and furnishes power for lights and other auxiliaries when the engine is shut down.

WHEELSLIP CORRECTION—Wheelslip is automatically detected by comparison of output signals from current measuring reactors located in each motor circuit. Slip is automatically corrected by automatic application of sand and reduction of power.

GROUND RELAY PROTECTION—If a ground occurs, engine speed returns to idle, power is removed, the alarm bell rings, and visual indication is given to the operator.

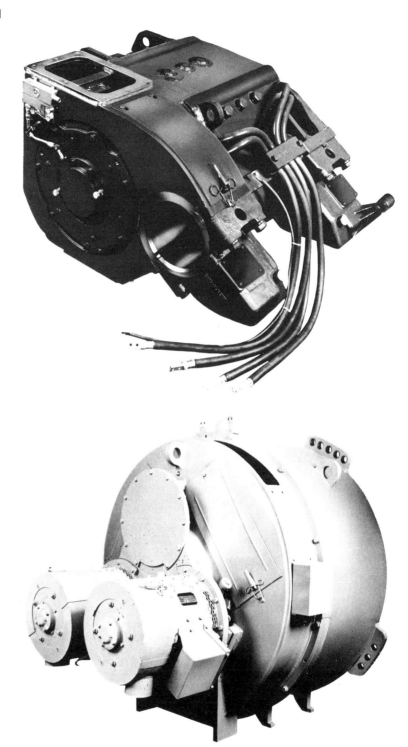

performance characteristics

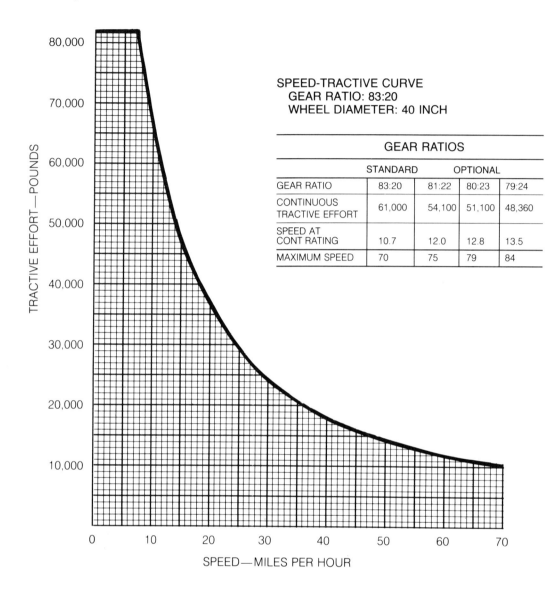

SPEED-TRACTIVE CURVE
  GEAR RATIO: 83:20
  WHEEL DIAMETER: 40 INCH

	GEAR RATIOS			
	STANDARD	OPTIONAL		
GEAR RATIO	83:20	81:22	80:23	79:24
CONTINUOUS TRACTIVE EFFORT	61,000	54,100	51,100	48,360
SPEED AT CONT RATING	10.7	12.0	12.8	13.5
MAXIMUM SPEED	70	75	79	84

General Electric Diesel-Electric Switching Locomotive Specifications

Reliable, Fuel-Efficient Performance For Yard And Branch Line Service

General Electric switching locomotives are designed to provide optimum horsepower and tractive effort in yard and branch line service. This optimum horsepower achieves excellent fuel economy.

The General Electric switching locomotive has two carefully selected, fuel-efficient, four-cycle turbocharged diesel engines.

General Electric's permanent parallel-connected traction motor concept provides 18 to 48 percent more usable tractive effort than equivalent road switchers. GE switchers have more pulling power, yet use up to 70 percent less fuel at idle than road switchers. Easy start capabilities encourage shutdown during idle periods, saving even more fuel.

These high-performance capabilities along with recent investments in state-of-the-art facilities, product development, and training technologies represents GE's dedication to provide unmatched quality and reliability.

Operator Safety And Comfort

General Electric's sound and heat-insulated operator's cab has been human engineered for operator safety and comfort, 360° visibility, and control efficiency.

From an adjustable, three-way comfort seat, the operator has easy access to an internally illuminated instrument and control panel.

The independent and automatic air brake system includes a deadman foot pedal safety control feature. The system features dual air compressors, each rated at a speed corresponding to full load engine speed.

FRA Regulation

The General Electric switching locomotive is designed and built in accordance with GE's interpretaton of applicable Federal Railway Administration (FRA) regulations.

OSHA Requirements

The locomotive is provided with features to facilitate compliance with U.S. Occupational Safety & Health Act (OSHA) requirements. Insulated operator's cab and muffled engines minimize engine exhaust noise. Engine air/fuel control valves control exhaust emissions. Safety devices such as the deadman switch, equipment guards, roughened walkways, and FRA high impact safety glass are also standard.

Materials

All materials are in accordance with standard material specifications of the General Electric Company. They compare (or are superior) to those recommended by the various engineering, testing and operating societies such as ASME, IEEE, ASTM, AAR, and ANSI.

Superstructure

The welded steel superstructure consists of an operator's cab and two engine hoods. The operator's cab has a non-metallic floor covering as well as heat and sound insulation on the roof and sides. Windows are in the sides and ends of the cab. Two of the windows in each side are sliding type, all others are fixed. All fixed windows have rubber sash, are heat resistant, and are glazed with FRA high impact safety glass.

Two steel doors (located on the left front and right rear of the cab) allow access from the walkways. A window is in each door.

Removable engine hoods are equipped with easy access doors. A formed steel bulkhead within the engine compartment separates engine and traction generator. Impingement filters in this generator compartment provide clean air to the generator, motors, and control compartment to help minimize maintenance.

Underframe

The rugged box section underframe is fabricated from electrically-welded rolled steel plates. Crossover platforms with recessed type FRA approved steps are on each end of the locomotive.

These steps are lighted and provide access to walkways on both sides of the locomotive. The frame is designed for buff loads of 2½ Gs with less than ½-inch deflection at center. Walkways and step treads have roughened surfaces. Step and walkway handrails are fastened to the outside of the deck.

Automatic AAR couplers and rubber draft gear are standard. A variety of hook and buffer arrangements and alternative coupler locations are available to meet specific requirements.

Controls

The standard locomotive has single-station, single-unit control equipment. The operator's station is on the right hand side of the operator's cab. A console is located in the front center of the cab. Conveniently grouped at this station are the throttle, reverser, and brake valve levers, sander, bell, horn, and window wiper controls, and the control switches. The gage panel is centrally located on the slanted portion of the console.

Direction of motion is selected by placing the reversing lever in the proper position while the throttle lever is in the idling position. Locomotive power is controlled by the throttle lever, which automatically regulates the output of each engine simultaneously and is interlocked with the electrical control. This interlocking establishes the electric circuit necessary for proper application of power to the traction motors. It also prevents reversal of the motors under power.

The pressurized control equipment compartment, located under the operator's cab, is serviced from the side of the locomotive. The locomotive is equipped with ground relay protection for traction equipment. Foot-operated safety (deadman) control is provided.

A duplicate set of operating controls and gage panels, opposite the single-unit control, is available as an option for customers who have dual station applications.

Brakes

The Schedule 26NL air brake system has both independent and automatic functions. Foot pedal operated safety (deadman) control is furnished. Two duplex brake cylinders, mounted on each truck, operate fully-equalized brake rigging which applies two shoes to each wheel. The brake shoe force equals approximately 60 percent of the weight on the wheels at a cylinder pressure of 50 PSI (3.52 kg per sq cm).

Two air-cooled, two-stage, belt-driven air compressors operate against a reservoir pressure of 130 lb PSI (914 kg per sq cm). Combined capacity of the main air reservoirs is approximately 43,000 cu in (705 liters).

A pump-action hand brake provides braking to hold the locomotive at a standstill.

A combination pneumatic/battery compartment is located under the operator's cab and can be serviced from the side. The battery compartment is ventilated and drained. Gasketed trap doors in the cab floor are provided for routine servicing from above.

Traction Generator

Each of the engines is directly connected to a direct-current shunt-wound generator. One end of the generator armature is supported by an anti-friction bearing mounted in the frame head. The other end is connected to the engine flywheel by a laminated steel disc flexible coupling. The magnet frame is bolted to the engine flywheel housing. The complete engine-generator set mounts on the locomotive platform.

The SL-80 and SL-110 engines are started by storage battery power, using the traction generators equipped with starting windings. The SL-144 engines are cranked with starter motors using storage battery power.

Two 60-ampere alternators, one mounted on each engine, furnish auxiliary power at a constant potential over the full operating range of engine speed. Alternator output is rectified and regulated. One alternator has sufficient capacity to supply all control and lighting loads, and charge the batteries.

Traction Motors

Two traction motors are connected in parallel to each generator. These dc series-wound motors drive through double reduction spur gearing. They are carried on the axle-mounted gear housing, which is supported from the truck frame by a resilient nose suspension. Motors are force-ventilated with one blower for each pair of motors.

A safety bar prevents the motor frame from rotating in the event of a motor suspension bolt failure.

Running Gear

The running gear consists of two 2-axle, spring over journal box, swivel trucks. This construction eliminates the need for side equalizers as well as the wear and maintenance associated with equalizer ends. It also simplifies truck component assembly and disassembly.

Sand boxes are integral with the truck frames for improved sand delivery to the rails. Flexible rubber sand hoses withstand derailment and obstructions along rail.

The journal boxes feature 6½ × 12-inch sealed cartridge-type roller bearings for minimum maintenance. Journal box guides and side bearings are equipped with renewable steel wear plates. The 22.5-inch diameter centerplate is oil lubricated and has seals to minimize contamination.

Wheels are heat treated solid rolled steel with AAR standard tread and flange contours. Axles are forged carbon steel.

General Information

Radiators

One-piece fan-blown radiators cool the engine jacket water. An automatically actuated fan disconnect clutch is provided.

This feature connects the radiator fan only when the water temperature rises above ideal engine operating temperature and radiator cooling is required.

Lube And Fuel Oil System

The lube oil filters consist of one full-flow type and one bypass-type per engine. Dry-type two-stage inertial intake air cleaners have replaceable second stage paper elements.

The fuel oil system consists of a baffled tank, eight reservoir gages, two filling connections, two drain and two clean-out plugs. Emergency shut-off is achieved from remote electrical control trips on each side of the locomotive and in the operator's cab.

A warning buzzer sounds if a low lube oil pressure condition exists. Low lube oil automatically shuts down the engine. A warning buzzer also sounds if high water temperature exists. Automatic power removal and return to high idle speed further protects the engines.

Two engine hour meters, one for each engine, assist in maintenance schedules.

Batteries

Four 8-volt, 4-cell lead acid batteries connected in series (167 amp/hr at eight-hour rate) provide engine starting power and auxiliary power when the engines are shut down. Auxiliary alternators charge these batteries at 36.5 volts.

Testing

All component parts are given standard commercial tests before locomotive assembly.

The locomotive is tested as follows:

1. Wiring is checked by observing operating sequence of electrical devices and by testing for continuity of circuit between terminals.
2. High potential tests of power and control circuits are made in accordance with ANSI standards.
3. Tests are made to check air pipe leakage, operation of air brakes, and other air-operated devices.
4. Meters and gages are calibration checked.
5. Power plants are tested to full load to check and adjust generator output and engine performance, including power and speed.

Each power plant is tested and adjusted to deliver full rated horsepower at full engine speed.

Painting

The interior of the locomotive is finished in gray. The exterior cab, hoods, and sides of the platform are painted in yellow - General Electric Number 113.

The platform top, underframe and running gear are painted in black - General Electric Number 101. End frames are finished in black and yellow chevron stripes at a 45° angle.

The customer's name and road number can be applied in black block letters on each side of the operator's cab.

Non-standard or multiple colors, stripes, insignias, and monograms can be applied at extra cost.

Standard Equipment

Arm Rest - Two.
Bell - One, non-swinging type, with air operated ringer.
Cab Heater - One electric fan blown, hot-water type.
Cab Lights - Two ceiling, plus individual gage lights.
Conductor's Valve - One, on helper's side.
Defrosters - Four.
Engine Exhaust Mufflers - Two (one per engine).
Field Shunting - Permits continued full utilization of available horsepower at locomotive speeds above 15 miles per hour (24 kph). Operation is completely automatic. Standard on the SL-144. Optional on SL-80 and SL-110.
Fire Extinguisher - One 20-pound dry chemical type suitable for Class A, B, and C fires. Located in main cab.
Fuel Fill - Two gravity filling connections.
Ground Relay - Power circuit ground protection with reset switch and indication light in the operator's cab.
Headlights - Two on each end, 200-watt sealed-beam type.
Horn - Two, one facing each direction, single trumpet type mounted on top of the engine cabs near the exhaust stacks.
Marker Lights - Eight, four per end (two red, two white).
Safety Control - Deadman foot-operated switch.
Sanders - Eight, air-operated.
Seat - One slide-mounted, swiveling, vertically adjustable, upholstered, with back rest.
Speedometer - One, electric.
Step Lights - Four, one at each access step.
Sun Visors - Four, fully adjustable.
Window Wipers - Four, air-operated.

Optional Features And Modifications

The following modifications may change locomotive weight, dimensions, and price:

*__Air-Operated Uncoupling Device__ - For remote uncoupling of cars from the locomotive.
*__Air Car Dump Lines__ - With or without operating valve in the operator's cab.
*__Air Conditioning__ - Two complete units with one compressor driven from each engine. Maximum total rating at full engine speed is 40,000 BTU/hour.
*__Awnings__ - Steel for the cab windows. On each side of the operating cab, if clearance diagram permits installation.
*__Ballast__ - Standard maximum weight for the SL-80 is 80 tons (72 metric tons). Ballast can be removed to meet any axle load requirement to minimum of 65 tons (59 metric tons). Standard maximum weight of SL-110 is 110 tons (99.8 metric tons). Ballast can be removed to meet any axle load requirement to a minimum of 85 tons (77.1 metric tons). Standard maximum weight for the SL-144 is 144 tons (130.6 metric tons). Ballast can be removed to meet any axle load requirement to a minimum of 115 tons (104.3 metric tons).
Battery Charging Receptacles - Mounted on each side of the locomotive to connect a wayside battery charger to the locomotive. Mating plugs included in spares.
Battery - Optional nickel-cadmium.
*__Bay Windows__ - For winter operation.
Brakes-1) Independent straight air option. 2) Straight air locomotive brakes and vacuum train brakes option. 3) Composite material brake shoe option.
Buffer Lights - For buffer or end frame area.
Cab Ventilating Fans - Fixed, mounted on ceiling in operator's cab.
*__Classification Lights__ - In place of marker lights.
*__Communications Radio__ - To meet regulations as stated in AAR Communications Manual Specification 12-10 and requirements of locomotive 37.5-volt system.
Coupler Arrangement - Alternate types to suit customer's requirements.
Coupler Height - To suit customer's requirements.
Dual Station Control.
End Step And Grab Irons - Where permitted.
Engine - Caterpillar engines.
Engine Water Heaters - Two, with receptacles and plug. (Not a substitute for antifreeze). Customer must specify desired AC voltage - 110, 220, or 440.
Extra Cab Heater.
Flange Lubricators - Four, graphite stick type.
Fuel Fill - Automatic rapid fueling connections are available (Buckeye or Houston).
High Ambient Cooling System - For 130° F.
Hubodometer - Axle-driven, for distance recording.
Locomotive Overspeed Protection - Warning whistle with automatic power removal and brake application after time delay.
Mirrors - Two mounted inside the operator's cab, or alternatively outside of the cab if clearance permits.
*__Multiple-Unit Control__ - To enable the operation of two or three similar locomotive units from one operator's cab.
Refrigerator - Ajax.
Reusable Car Body Filters.
Seat - Additional, mounted on the left side of the operator's cab.
Snow Plow - 36 inches (either or both ends).
Snubbers - (Standard on SL-144)
Speed Control - Automatic slow speed for dump or load operations. Range - Low (.25 to 1.0 mph); medium (.75 to 3.0 mph); high (2.0 to 7.0 mph).
Speed Recorder - Combination speedometer and speed recorder mounted in the engine compartment. Customer to specify mph or kph scale.
Steel-Tired Wheels - Steel tires mounted on rolled steel centers.
Warning Lights - High intensity strobe light mounted on the center of the cab roof.
Wheelslip Control - Detects and corrects wheel slippage by automatic sanding and application of brakes.

*Check General Electric for compatibility when desiring two or more items marked with an asterisk.

Location Of Apparatus

1. Jacking And Lifting Lugs
2. Traction Motor And Gear Box
3. Air Reservoir
4. Fuel Tank
5. Panel Air Filters
6. Pressurized Electrical Control Compartment*
7. Three-Way Adjustable Seat
8. Control Console
9. Sound Insulation
10. Diesel Fuel Fill
11. Air Compressor
12. Traction Motor Blower
13. Engine Compartment Bulkhead
14. Traction Generator
15. Four-Cycle Turbocharged Diesel Engine
16. Radiator
17. Engine Exhaust Pipe
18. Exhaust Muffler
19. Inertial Engine Air Filter
20. Marker Lights (Optional Class Lights) And Dual Headlights
21. Automatic Fan Clutch
22. Truck With Springs Over Journals, Clasp Brakes, 4-Wheel Sanding, Sealed Roller Bearings
23. Cabling Behind Side Plate, Piping On Opposite Side
24. Sand Box Fill
25. FRA Steps And Step Lights
26. Non-Skid Walkway
27. Couplers To Suit

*Actually located on the left side of the locomotive. Displayed for illustration purposes only.

SL-80 Diesel-Electric Switching Locomotive
Specification RY24781B

	Haulage-Ability Trailing Tons Locomotive Can Handle U.S. TONS				
SPEED MPH		4	Continuous Broad Gage 6	10	15
Level	65T	---	5935	3735	1635
	80T	7920	5920	3720	1620
1% GRADE	65T	---	1135	695	275
	80T	1520	1120	680	260
2% GRADE	65T	---	602	357	124
	80T	809	587	342	109
3% GRADE	65T	---	397	227	66
	80T	535	382	212	51

	Haulage-Ability Trailing Tons Locomotive Can Handle METRIC TONS				
SPEED KmPH		6.4	Continuous Broad Gage 10	16	24
Level	59MT	---	5395	3395	1486
	72MT	7200	5382	3382	1473
1% GRADE	59MT	---	1032	632	250
	72MT	1382	1018	618	236
2% GRADE	59MT	---	547	325	113
	72MT	735	534	311	99
3% GRADE	59MT	---	361	206	60
	72MT	486	347	193	46

NOTE: Rolling resistance assumed to be 5 lb/ton.

SL-80 Diesel-Electric Switching Locomotive
Specification RY24781B

Ratings And Performance Features

Equipment Summary
Two Diesel Engines	Cummins NT-885L4
Two Traction Generators	GT-558
Four Traction Motors	GE-763
Two Compressors	100 cfm (2832 liters/min) each

Locomotive Specifications
Weight	65-80 tons (59-72 metric tons)
	Manufacturing tolerance ±2%
Maximum Permissible Speed	
Narrow Gage	26 mph (42 km/hr)
Broad Gage	21 mph (34 km/hr)
Hp Input To Generators	600 hp
Tractive Effort (Starting)	30% adhesion
65 Tons	39,000 lb (17,690 kg)
80 Tons	48,000 lb (21,818 kg)
Tractive Effort (Cont.)	
Broad Gage	31,600 lb (14,311 kg)
Narrow Gage	25,600 lb (11,612 kg)

Dimensions And Capacities
Gage	36-66 in (914-1676 mm)
Max. Height Above Rail	12 ft 10.5 in (3924 mm)
Min. Clearance Above Rail	3-3/4 in (95 mm)
Length Over End Frames	38 ft (11583 mm)
Truck Distance C. To C.	20 ft (6096 mm)
Truck Wheel Base	7 ft. 9 in (2362 mm)
Max. Width Overall	9 ft. 6 in (2896 mm)
Fuel Oil Capacity, Total	400 gal (1514 liters)
Lube Oil Capacity, Total	23 gal (87 liters)
Water Capacity, Total	30 gal (144 liters)
Sand Capacity (Eight Wheels)	16 cu ft (453 liters)
Outline Drawing	41D722658
Min. Curve Radius	
(Locomotive Alone)	75 ft (22.9 m)
(Multiple Unit)	100 ft (30.5 m)

Engines
Model (Two)	Cummins NT855L4
Horsepower (Each Engine)	335/300
Peak Torque (1500 rpm)	1006 lb ft (139 Kgf m)
Number Of Cylinders	6
Cylinder Arrangement	In-Line
Compression Ratio	14.3:1
Nominal Torque Rise	20%
Stroke Cycle	4
Bore	5½ in (139.7 mm)
Stroke	6 in (152.4 mm)
Full Speed/Rated Speed	2100 rpm
Idle Speed	650 rpm
Piston Speed	2100 ft/min (640 m/min)
Aspiration	Turbocharged
Displacement	855 cu in (14.0 liters)
Oil System Capacity	8.6 U.S. gal (32.6 liters)
Coolant Capacity	5 U.S. gal (18.9 liters)
Net Dry Weight	2770 lb (1256 Kg)

Electric Drive System

GT-558 Generator
Weight	1860 lb (844 Kg)
Rated Voltage (Nominal)	300 volts
Max. Speed	2100 rpm
Hp Input For Traction	300
Amperes (Cont.)	820 amps
Amperes (1 Hr.)	880 amps
Ventilation	self

GE-763 Traction Motors
Weight	1135 lb (515 Kg)
Rated Voltage	300 volts
Max. Speed	4500 rpm
Hp Input For Traction	210
Amperes (Cont.)	410 amps
Amperes (1 Hr.)	435 amps
Torque (Cont.)	520 lb ft (71.8 Kgf m)
Torque (1 Hr.)	565 lb ft (78 Kgf m)
Forced Ventilation	500 cfm at
	35.56 mm H O
	1.4 inches H O

GA-70 Gear Unit
Motor Application	GE-763
Gear Ratio	20.9:1
Wheel Diameter	33 in (838 mm)
Optional Wheel Diameter	36 in (914 mm)
Tractive Effort (Cont.) With Motor	7900 lb (3583 Kg)
Tractive Effort (1 Hr.) With Motor	8590 lb (3896 Kg)
Weight (Gear Unit Only)	1648 lb (748 Kg)
Track Gage (Min.)	56.5 in (1435 mm)
Max. Speed	21 mph (34 km/hr)

Air Systems

Compressors
Two At 100 cfm (2832 liters/min) each
Air Cooled
Two Stage
Belt Driven

Brakes
Schedule 26NL
Independent And Automatic Functions
Foot Pedal Operated Safety Deadman Control
Two Duplex Brake Cylinders Per Truck
Clasp Brakes
Equalized Brake Rigging
Brake Shoe Force Equals Approx. 60% Of Weight On The Wheels At 50 psi (3.52 kg per sq cm)

Air Reservoirs
Combined Capacity Of Main Reservoirs
43,000 cu in (705 liters)
Pressure Rated For 130 psi (914 kg per sq cm)

Park Brake
Pump Action Hand

Domestic Shipping Information

Dimensions

	Length (over end plates)	38 ft
	Width	9 ft 6 in
	Height	12 ft 10.5 in
	Truck Centers	20 ft
	Axle Centers	7 ft 9 in
Weight	Short Tons	62 to 77

Export Shipping Information

Dimensions
(with boxing)

Carbody	Length	11,989 mm
	Width	2,946 mm
	Height	3,912 mm
Truck Assembly (each)	Length	4,013 mm
	Width	2,591 mm
	Height	1,219 mm

Weights

Carbody	Kilograms	35,060 - 48,660
Truck Assembly (each)	Kilograms	13,730

Standard Equipment

Arm Rest	Headlights
Bell	Horn
Cab Heater	Marker Lights
Cab Lights	Safety Control
Conductor's Valve	Sanders
Defrosters	Seat
Engine Exhaust Mufflers	Speedometer
Fire Extinguisher	Step Lights
Fuel Fill	Sun Visors
Ground Relay	Window Wipers

Optional Equipment

Air-Operated Uncoupling Device	Field Shunting
Air Car Dump Lines	Flange Lubricators
Air Conditioning	Fuel Fill
Awnings	High Ambient Cooling System
Ballast	Hubodometer
Battery Charging Receptacle And Plug	Locomotive Overspeed Protection
Battery	Mirrors
Bay Windows	Multiple-Unit Control
Buffer Lights	Radio Remote Control
Cab Ventilating Fans	Refrigerator
Classification Lights	Reusable Car Body Filters
Communications Radio	Snow Plow
Coupler Arrangement	Speed Control
Coupler Height	Speed Range
Dual Station Control	Speed Recorder
End Step And Grab Irons	Steel-Tired Wheels
Engine Water Heaters	Warning Lights
Extra Cab Heater	Wheelslip Control

(Weights Indicated Vary In Proportion To Ballast Added).

CONVERSION TABLE

System of Measure:

U.S.A.	C.G.S.	*SI
1 in	= 25.4 mm	= 25.4 mm
1 sq in	= 6.452 cm^2	= 6.452 cm^2
1 cu in	= 16.387 cm^3	= 16.387 cm^3
1 ft	= 0.3048 m	= 0.3048 m
1 cu ft	= 0.02832 m^3	= 0.02832 m^3
1 cu yd	= 0.7646 m^3	= 0.7646 m^3
1 lb	= 0.4536 kg	= 0.4536 kg
1 gal	= 3.785 lit	= 3785 cm^3
1 GPM	= 3.785 lit/min	= 63.09 cm^3/s
1 cfm	= 28.32 lit/min	= 472.00 cm^3/s
1 psi	= .0703 kg/cm^2	= 6.895 KPa
1 lb-ft	= 0.138 kgf m	= 1.356 Nm
1 hp	= 1.014 metric hp	= 0.746 KW
1 mph	= 1.609 km/h	= 1.609 km/h
1 ton (short)	= 0.907 metric ton	= 907 kg

*International System of Units (Systeme International)

SL-80 Dimensional Drawings
41D722658

Specification - RY24781B

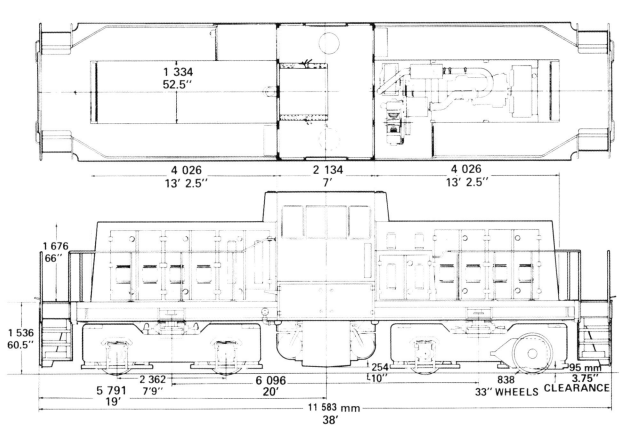
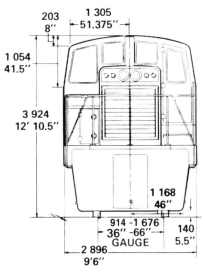

SL-80 Major Components

Model GT-558 Traction Generator

Model GE-763 Traction Motor With Gear Unit/Wheel/Axle Assembly

Cummins NT-855L4 Diesel Engine

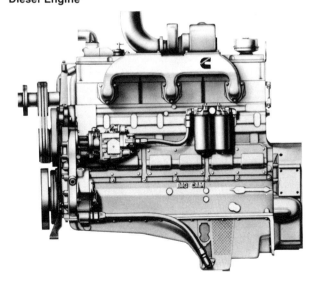

SL-80 Major Component Drawings

GT-558 Traction Generator Dimensional Drawing

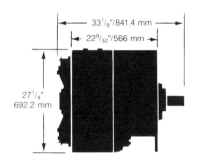

GE-763 Traction Motor And GA-70 Gear Unit Dimensional Drawing

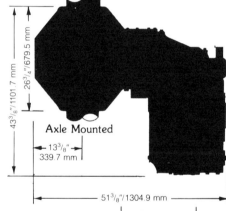

GE-763 Traction Motor Dimensional Drawing

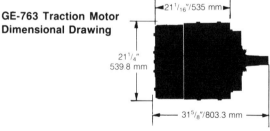

Engine Dimensional Drawing

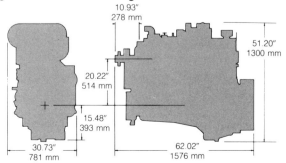

SL-80 Diesel-Electric Switching Locomotive
Specification RY24781B

Performance Characteristics

SL-80 SWITCHING LOCOMOTIVE AT 600 HORSEPOWER

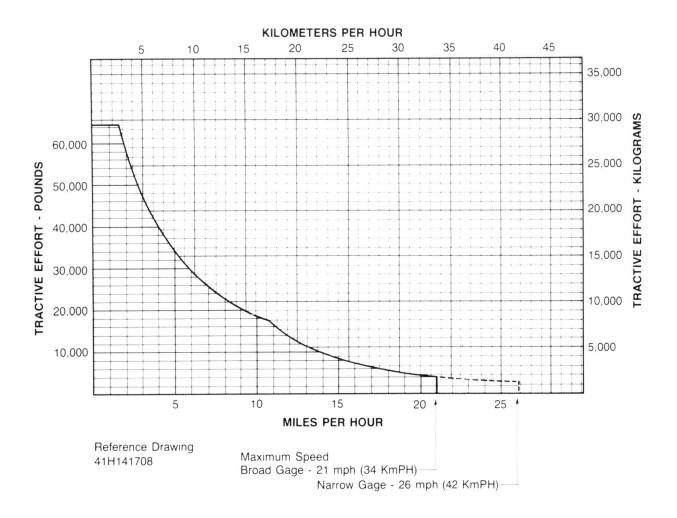

Reference Drawing
41H141708

Maximum Speed
Broad Gage - 21 mph (34 KmPH)
Narrow Gage - 26 mph (42 KmPH)

SL-110 Diesel-Electric Switching Locomotive
Specification RY24782B

Ratings And Performance Features

Equipment Summary

Two Diesel Engines	Cummins NT-885L4
Two Traction Generators	GT-558
Four Traction Motors	GE-763
Two Compressors	100 cfm (2832 liters/min) each

Locomotive Specifications

Weight	85-110 tons (77.1-99.8 metric tons) Manufacturing tolerance ±2%
Maximum Permissible Speed	21 mph (34 km/hr)
Hp Input To Generators For Traction	600 hp (609 cv)
Wheel Diameter	33 in (838 mm)
Tractive Effort (Starting)	30% adhesion
85 Tons	51,000 lb (23,133 kg)
110 Tons	66,000 lb (29,937 kg)
Tractive Effort (Cont.)	31,600 lb (14,331 kg)
Narrow Gage	25,600 lb (11,612 kg)

Dimensions And Capacities

Gage	56½-60 in (1435-1676 mm)
Max. Height Above Rail	12 ft 10.5 in (3924 mm)
Min. Clearance Above Rail	3-3/4 in (95 mm)
Length Over End Frames	41 ft (12497 mm)
Truck Distance C. To C.	22 ft (6705 mm)
Truck Wheel Base	7 ft 9 in (2362 mm)
Max. Width Overall	9 ft 6 in (2896 mm)
Fuel Oil Capacity, Total	500 gal (1893 liters)
Lube Oil Capacity, Total	23 gal (87 liters)
Water Capacity, Total	30 gal (144 liters)
Sand Capacity (Eight Wheels)	16 cu ft (453 liters)
Outline Drawings	41D722659 41D760834
Min. Curve Radius	
(Locomotive Alone)	75 ft (22.9 m)
(Multiple Unit)	100 ft (30.5 m)

Engines

Model (Two)	Cummins NT855L4
Horsepower (Each Engine)	335/330
Rated Speed	2100 rpm
Peak Torque (1500 rpm)	1006 lb ft (139 Kgf m)
Number Of Cylinders	6
Cylinder Arrangement	In-Line
Compression Ratio	14.3:1
Nominal Torque Rise	20%
Stroke Cycle	4
Bore	5½ in (139.7 mm)
Stroke	6 in (152.4 mm)
Full Speed	2100 rpm
Idle Speed	650 rpm
Piston Speed	2100 ft/min (640 m/min)
Aspiration	Turbocharged
Displacement	855 cu in (14.0 liters)
Oil System Capacity	8.6 U.S. gal (32.6 liters)
Coolant Capacity	5 U.S. gal (18.9 liters)
Net Dry Weight	2770 lb (1256 Kg)

Electric Drive System

GT-558 Generator

Weight	1860 lb (844 Kg)
Rated Voltage (Nominal)	300 volts
Max. Speed	2100 rpm
Hp Input For Traction	300
Amperes (Cont.)	820 amps
Amperes (1 Hr.)	880 amps
Ventilation	self

GE-763 Traction Motors

Weight	1135 lb (515 Kg)
Rated Voltage	300 volts
Max. Speed	4500 rpm
Hp Input For Traction	210
Amperes (Cont.)	410 amps
Amperes (1 Hr.)	435 amps
Torque (Cont.)	520 lb ft (71.8 Kgf m)
Torque (1 Hr.)	565 lb ft (78 Kgf m)
Forced Ventilation	500 cfm at 1.4 inches H₂O 35.56 mm H₂O

GA-70 Gear Unit

Motor Application	GE-763
Gear Ratio	20.9/1
Wheel Diameter	33 in (838 mm)
Tractive Effort (Cont.) With Motor	7900 lb (3583 Kg)
Tractive Effort (1 Hr.) With Motor	8590 lb (3896 Kg)
Weight (Gear Unit Only)	1648 lb (748 Kg)
Track Gage (Min.)	56.5 in (1435 mm)
Max. Speed	21 mph (34 km/hr)

Air Systems

Compressors

Two At 100 cfm (2832 liters/min) each
Air Cooled
Two Stage
Belt Driven

Brakes

Schedule 26NL
Independent And Automatic Functions
Foot Pedal Operated Safety Deadman Control
Two Duplex Brake Cylinders Per Truck
Clasp Brakes
Equalized Brake Rigging
Brake Shoe Force Equals Approx. 60% Of Weight On The Wheels At 50 psi (3.52 kg per sq cm)

Air Reservoirs

Combined Capacity Of Main Reservoirs
43,000 cu in (705 liters)
Pressure Rated for 130 psi (914 kg per sq cm)

Park Brake

Pump Action Hand

SL-110 Dimensional Drawings

Specification - RY24782B

41D722659   41D760834

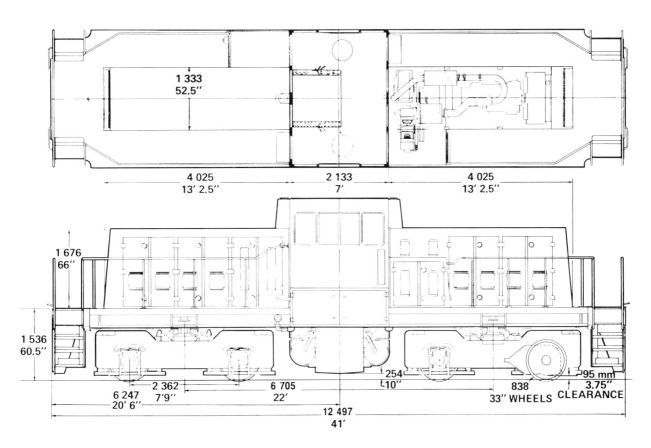
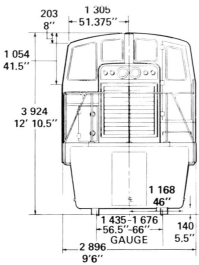

SL-110 Major Components

Model GT-558 Traction Generator

Model GE-763 Traction Motor With Gear Unit/Wheel/Axle Assembly

Cummins NT-855L4 Diesel Engine

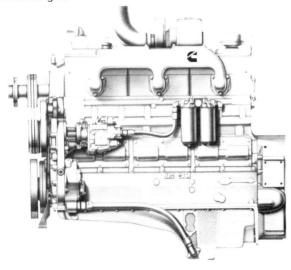

SL-110 Major Component Drawings

GT-558 Traction Generator Dimensional Drawing

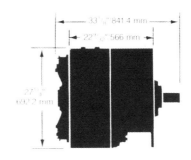

GE-763 Traction Motor And GA-70 Gear Unit Dimensional Drawing

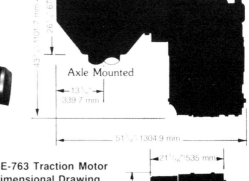

GE-763 Traction Motor Dimensional Drawing

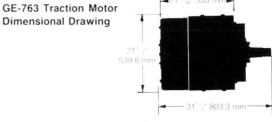

Engine Dimensional Drawing

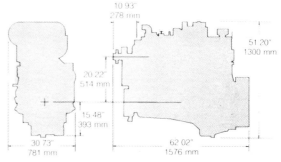

Domestic Shipping Information

Dimensions

	Length (over end plates)	41 ft
	Width	9 ft 6 in
	Height	12 ft 10.5 in
	Truck Centers	22 ft
	Axle Centers	7 ft 9 in
Weight	Short Tons	82 to 107

Export Shipping Information

Dimensions
(with boxing)

Carbody	Length	12,979 mm
	Width	2,946 mm
	Height	3,912 mm
Truck Assembly (each)	Length	4,013 mm
	Width	2,591 mm
	Height	1,219 mm

Weights

Carbody	Kilograms	52,934 - 69,218
Truck Assembly For 85-ton Locomotive	Kilograms	13,812
Truck Assembly For 110-ton Locomotive	Kilograms	17,264

(Weights Indicated Vary In Proportion To Ballast Added).

Standard Equipment

Arm Rest
Bell
Cab Heater
Cab Lights
Conductor's Valve
Defrosters
Engine Exhaust Mufflers
Fire Extinguisher
Fuel Fill
Ground Relay

Headlights
Horn
Marker Lights
Safety Control
Sanders
Seat
Speedometer
Step Lights
Sun Visors
Window Wipers

Optional Equipment

Air-Operated Uncoupling Device
Air Car Dump Lines
Air Conditioning
Awnings
Ballast
Battery Charging Receptacle And Plug
Battery
Bay Windows
Buffer Lights
Cab Ventilating Fans
Classification Lights
Communications Radio
Coupler Arrangement
Coupler Height
Dual Station Control
End Step And Grab Irons
Engine Water Heaters
Extra Cab Heater

Field Shunting
Flange Lubricators
Fuel Fill
High Ambient Cooling System
Hubodometer
Locomotive Overspeed Protection
Mirrors
Multiple-Unit Control
Narrow Gage to 100 Tons
Radio Remote Control
Refrigerator
Reusable Car Body Filters
Snow Plow
Speed Control
Speed Range
Speed Recorder
Steel-Tired Wheels
Warning Lights
Wheelslip Control

CONVERSION TABLE

System of Measure:

U.S.A.	C.G.S.	*SI
1 in	= 25.4 mm	= 25.4 mm
1 sq in	= 6.452 cm^2	= 6.452 cm^2
1 cu in	= 16.387 cm^3	= 16.387 cm^3
1 ft	= 0.3048 m	= 0.3048 m
1 cu ft	= 0.02832 m^3	= 0.02832 m^3
1 cu yd	= 0.7646 m^3	= 0.7646 m^3
1 lb	= 0.4536 kg	= 0.4536 kg
1 gal	= 3.785 lit	= 3785 cm^3
1 GPM	= 3.785 lit/min	= 63.09 cm^3/s
1 cfm	= 28.32 lit/min	= 472.00 cm^3/s
1 psi	= .0703 kg/cm^2	= 6.895 KPa
1 lb-ft	= 0.138 kgf m	= 1.356 Nm
1 hp	= 1.014 metric hp	= 0.746 KW
1 mph	= 1.609 km/h	= 1.609 km/h
1 ton (short)	= 0.907 metric ton	= 907 kg

*International System of Units (Systeme International)

SL-110 Diesel-Electric Switching Locomotive
Specification RY24782B

Haulage-Ability Trailing Tons Locomotive Can Handle U.S. TONS					
SPEED MPH		2.4	Continuous 6	10	15
Level	85T	---	5915	3715	1615
	110T	10890	5890	3690	1590
1% GRADE	85T	2115	1115	675	255
	110T	2090	1090	650	230
2% GRADE	85T	1137	582	337	104
	110T	1112	557	312	79
3% GRADE	85T	761	377	207	46
	110T	736	352	182	21

Haulage-Ability Trailing Tons Locomotive Can Handle METRIC TONS					
SPEED KmPH		3.9	Continuous 10	16	24
Level	77MT	---	5377	3377	1468
	99MT	9900	5354	3355	1445
1% GRADE	77MT	1923	1014	614	232
	99MT	1900	991	591	209
2% GRADE	77MT	1034	529	307	95
	99MT	1011	506	284	72
3% GRADE	77MT	692	343	188	42
	99MT	669	320	165	19

NOTE: Rolling resistance assumed to be 5 lb/ton.

SL-110 Diesel-Electric Switching Locomotive
Specification RY24782B

Performance Characteristics

SL-110 SWITCHING LOCOMOTIVE AT 600 HORSEPOWER

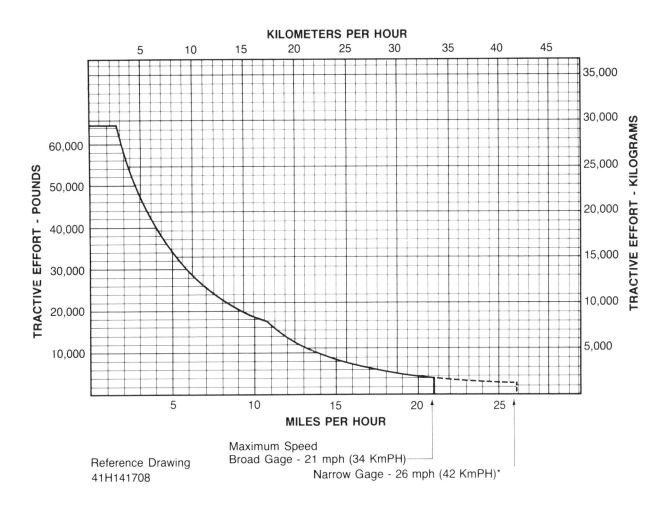

Reference Drawing 41H141708

Maximum Speed
Broad Gage - 21 mph (34 KmPH)
Narrow Gage - 26 mph (42 KmPH)*

*Narrow gage applicable up to 100 U.S. ton/91 metric ton locomotives depending on gear box model used.

SL-144 Diesel-Electric Switching Locomotive
Specification RY24784C

Ratings And Performance Features

Equipment Summary

Two Diesel Engines	KTA-1150L or optional KT-1150L
Two Traction Generators	GT-603
Four Traction Motors	GE-761
Two Compressors	143 cfm (4050 liters/min) each

Locomotive Specifications

Weight	115-144 tons (104.3-130.6 metric tons) Manufacturing tolerance ±2%
Maximum Permissible Speed	35 mph (56 km/hr)
Hp Input To Generator	1100 hp (1116 cv) or optional 800 hp (809 cv)
Wheel Diameter	40 in (1016 mm)
Tractive Effort (Starting)	30% adhesion
115 Tons	69,000 lb (32,970 kg)
144 Tons	86,400 lb (39,191 kg)
Tractive Effort (Cont.)	61,560 lb (27,920 kg)

Dimensions And Capacities

Gage	56½-66 in (1435-1676 mm)
Max. Height Above Rail	13 ft 3 in (4039 mm)
Min. Clearance Above Rail	5.5 in (140 mm)
Length Over End Frames	45 ft (13716 mm)
Truck Distance C. To C.	25 ft (7620 mm)
Truck Wheel Base	8 ft 6 in (2591 mm)
Max. Width Overall	9 ft 6 in (2896 mm)
Fuel Oil Capacity, Total	700 gal (2650 liters)
Lube Oil Capacity, Total	39 gal (148 liters)
Water Capacity, Total	40 gal (151 liters)
Sand Capacity (Eight Wheels)	16 cu ft (453 liters)
Outline Drawings	41D722781
Min. Curve Radius	
(Locomotive Alone)	110 ft (30.5 m)
(Multiple Unit)	150 ft (45.7 m)

Engines

Engines (Two)	KTA-1150L (600/550 hp per engine) or optional KT-1150L (450/400 hp per engine)
Rated Speed	2100 rpm
Peak Torque (1600 rpm)	1650 lb ft (228 Kgf m)
Torque Rise	10%
Number Of Cylinders	6
Cylinder Arrangement	In-Line
Stroke Cycle	4
Bore	6¼ in (153 mm)
Stroke	6¼ in (153 mm)
Full Speed	2100 rpm
Idle Speed	650 rpm
Piston Speed	2180 ft/min (640 m/min)
Aspiration KTA-1150L	Turbocharged/aftercooled
Aspiration KT-1150L	Turbocharged only
Displacement	1150 cu in (18.8 liters)
Oil System Capacity	12.5 U.S. gal (47.3 liters)
Coolant Capacity	8 U.S. gal (30.3 liters)
Net Dry Weight	3725 lb (1690 Kg)

Electric Drive System

GT-603 Generator

Weight	3900 lb (1769 Kg)
Rated Voltage (Nominal)	600 volts
Max. Speed	2100 rpm
Hp Input For Traction	1000
Amperes (Cont.)	1320 amps
Amperes (1 Hr.)	1350 amps
Ventilation	self

GE-761 Traction Motors

Weight	3460 lb (1569 Kg)
Rated Voltage	600 volts
Max. Speed	3136 rpm
Hp Input For Traction	450
Amperes (Cont.)	645 amps
Amperes (1 Hr.)	655 amps
Torque (Cont.)	2575 lb ft (355 Kgf m)
Torque (1 Hr.)	2627 lb ft (362.5 Kgf m)
Forced Ventilation	1680 cfm at 4.25 inches H₂O 47.58 cm m at 107.95 mm H₂O

GA-72 Gear Unit

Motor Application	GE-761
Gear Ratio	10.52/1
Wheel Diameter	40 in (1016 mm)
Tractive Effort (Cont.) With Motor	16,255 lb (7360 Kg)
Tractive Effort (1 Hr.) With Motor	16,580 lb (7521 Kg)
Weight (Gear Unit Only)	2130 lb (966 Kg)
Track Gage (Min.)	56.5 in (1435 mm)
Max. Speed	35 mph (56 km/hr)

Air Systems

Compressors

Two At 143 cfm (4050 liters/min) each
Air Cooled
Two Stage
Belt Driven

Brakes

Schedule 26NL
Independent And Automatic Functions
Foot Pedal Operated Safety Deadman Control
Two Duplex Brake Cylinders Per Truck
Clasp Brakes
Equalized Brake Rigging
Brake Shoe Force Equals Approx. 60% Of Weight
On The Wheels At 50 psi (3.52 kg per sq cm)

Air Reservoirs

Combined Capacity Of Main Reservoirs
43,000 cu in (705 liters)
Pressure Rated For 130 lb per sq in (914 kg per sq cm)

Park Brake

Pump Action Hand

SL-144 Dimensional Drawings
41D722781

Specification - RY24784C

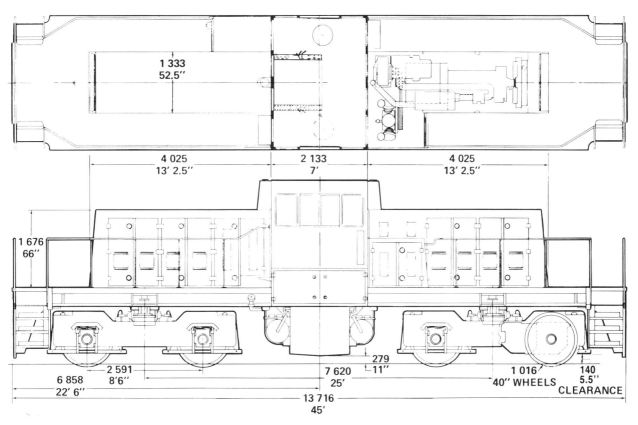
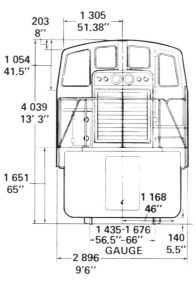

-149-

SL-144 Major Components

Model GT-603 Traction Generator

GE-761 Traction Motor And
General Electric GA-72 Gear Unit

Cummins KT/KTA-1150 Diesel Engine

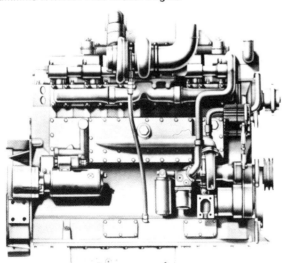

SL-144 Major Component Drawings

GT-603 Traction Generator
Dimensional Drawing

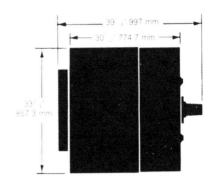

GE-761 Traction Motor And GA-72 Gear Unit
Dimensional Drawing

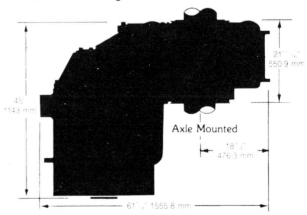

Cummins KT/KTA-1150 Diesel Engine
Dimensional Drawing

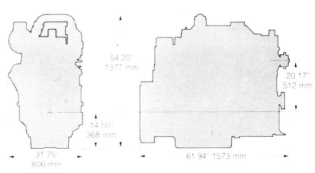

Domestic Shipping Information

Dimensions

	Length (over end plates)	45 ft.
	Width	9 ft 6 in
	Height	13 ft 3 in
	Truck Centers	25 ft
	Axle Centers	8 ft 6 in
Weight	Short Tons	111.5 to 140.5

Export Shipping Information

Dimensions
(with boxing)

Carbody	Length	14,173 mm
	Width	2,946 mm
	Height	4,039 mm
Truck Assembly (each)	Length	4,572 mm
	Width	2,743 mm
	Height	1,321 mm

Weights

Carbody	Kilograms	75,692 - 89,751
Truck Assembly (each)	Kilograms	21,818

Standard Equipment

Arm Rest
Bell
Cab Heater
Cab Lights
Conductor's Valve
Defrosters
Engine Exhaust Mufflers
Fire Extinguisher
Fuel Fill
Ground Relay

Headlights
Horn
Marker Lights
Safety Control
Sanders
Seat
Speedometer
Step Lights
Sun Visors
Window Wipers

Optional Equipment

Air-Operated Uncoupling
 Device
Air Car Dump Lines
Air Conditioning
Awnings
Ballast
Battery Charging
 Receptacle And Plug
Battery
Bay Windows
Buffer Lights
Cab Ventilating Fans
Classification Lights
Communications Radio
Coupler Arrangement
Coupler Height
Dual Station Control
End Step And Grab Irons
Engine Water Heaters
Extra Cab Heater

Flange Lubricators
Fuel Fill
High Ambient
 Cooling System
Hubodometer
Locomotive Overspeed
 Protection
Mirrors
Multiple-Unit Control
Radio Remote Control
Refrigerator
Reusable Car Body Filters
Snow Plow
Speed Control
Speed Range
Speed Recorder
Steel-Tired Wheels
Warning Lights
Wheelslip Control

CONVERSION TABLE

System of Measure:

U.S.A.	C.G.S.	*SI
1 in	= 25.4 mm	= 25.4 mm
1 sq in	= 6.452 cm^2	= 6.452 cm^2
1 cu in	= 16.387 cm^3	= 16.387 cm^3
1 ft	= 0.3048 m	= 0.3048 m
1 cu ft	= 0.02832 m^3	= 0.02832 m^3
1 cu yd	= 0.7646 m^3	= 0.7646 m^3
1 lb	= 0.4536 kg	= 0.4536 kg
1 gal	= 3.785 lit	= 3785 cm^3
1 GPM	= 3.785 lit/min	= 63.09 cm^3/s
1 cfm	= 28.32 lit/min	= 472.00 cm^3/s
1 psi	= .0703 kg/cm^2	= 6.895 KPa
1 lb-ft	= 0.138 kgf m	= 1.356 Nm
1 hp	= 1.014 metric hp	= 0.746 KW
1 mph	= 1.609 km/h	= 1.609 km/h
1 ton (short)	= 0.907 metric ton	= 907 kg

*International System of Units (Systeme International)

(Weights Indicated Vary In Proportion To Ballast Added).

SL-144 Diesel-Electric Switching Locomotive
Specification RY24784C

Haulage-Ability Trailing Tons Locomotive Can Handle U.S. TONS				
SPEED MPH	4	Continuous 5.5	10	15
Level				
115T	---	---	7285	4685
125T	---	12187	7275	4675
144T	15056	12168	7256	4656
1% GRADE				
115T	---	2347	1365	845
125T	2915	2337	1355	835
144T	2896	2318	1336	816
2% GRADE				
115T	---	1253	707	418
125T	1564	1243	697	408
144T	1545	1224	678	389
3% GRADE				
115T	---	832	454	254
125T	1044	822	444	244
144T	1025	803	425	225

Haulage-Ability Trailing Tons Locomotive Can Handle METRIC TONS				
SPEED KmPH	6.4	Continuous 8.8	16	24
Level				
104MT	---	---	6623	4259
114MT	---	11079	6614	4250
131MT	13687	11062	6596	4233
1% GRADE				
104MT	---	2134	1241	768
114MT	2650	2125	1232	759
131MT	2633	2107	1215	742
2% GRADE				
104MT	---	1139	643	380
114MT	1422	1130	634	371
131MT	1405	1113	616	354
3% GRADE				
104MT	---	756	413	231
114MT	949	747	404	222
131MT	932	730	386	205

NOTE: Rolling resistance assumed to be 5 lb/ton.

SL-144 Diesel-Electric Switching Locomotive
Specification RY24784C

Performance Characteristics

1100 HP

Reference Drawing
41H141704

800 HP

Reference Drawing
41H141706

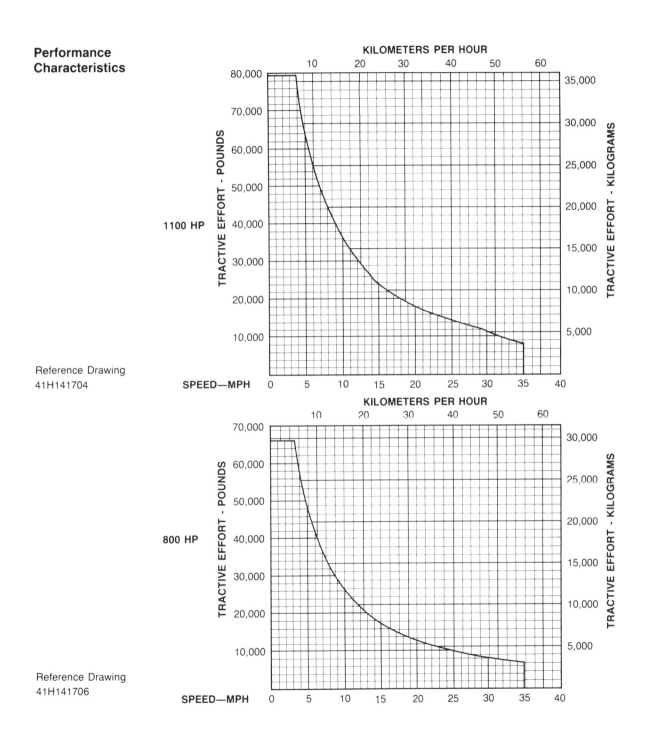

-153-

Ratings

	TRACTION GENERATORS			TRACTION MOTORS AND GEAR CASES	
MODEL NUMBER	GT-1519	GT-558	GT-603	GE-763	GE-761
Specification Number	RY24405	RY24468	RY24808	RY24552 (GA70) RY-24409 (GA41)	RY24809
Weight	1080 lb.	1860 lb.	3900 lb.	1135 lb.	3460 lb.
Rated Voltage (Nominal)	300 volts	300 volts	600 volts	300 volts	600 volts
Maximum Speed	2100 r.p.m.	2100 r.p.m.	2100 r.p.m.	4500 r.p.m.	3136 r.p.m.
Engine Horsepower Input For Traction (Maximum)	210	300	1000	210	450
Amperes (Continuous)	430 amps.	820 amps.	1320 amps.	410 amps.	645 amps.
Amperes (One Hour)	450 amps.	880 amps.	1350 amps.	435 amps.	655 amps.
Torque-Continuous (lb.-ft.)	---	---	---	520 lb.-ft.	2575 lb.-ft.
Torque-One Hour (lb.-ft.)	---	---	---	565 lb.-ft.	2627 lb.-ft.
Ventilation (Motor Commutator Chamber)	self	self	self	self or 500 c.f.m. at 1.4 in. H_2O	1680 c.f.m. at 4.25 in. H_2O
Outline	T-8855949	T-6751290	41E903367	41C635658	

GEAR UNIT				GA-41	GA-70	GA-72
Motor Application				GE-763	GE-763	GE-761
Gear Ratio				17.03/1	20.9/1	10.52/1
Wheel Diameter				33 in.	33 in.	40 in.
Tractive Effort (Cont.) With Motor				6400 lb.	7900 lb.	16,255 lb.
Tractive Effort (1 Hr.) With Motor				7000 lb.	8590 lb.	16,580 lb.
Weight (Gear Unit Only)				1900 lb.	1648 lb.	2130 lb.
Track Gage (Minimum)				36 in.	56.5 in.	56.5 in.
Maximum Speed				26 m.p.h.	21 m.p.h.	35 m.p.h.
Outline (Includes Motor)				41D732172	41D732175	41D731837

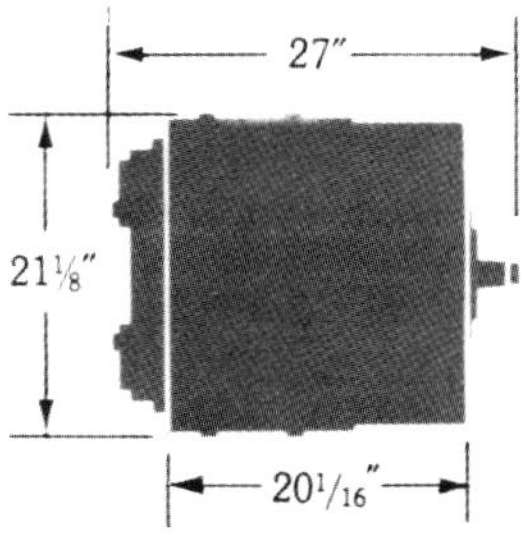

GT-1519 Traction Generator

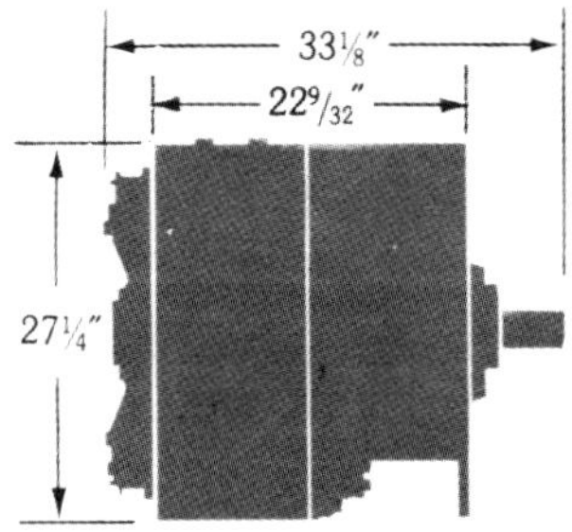

GT-558 Traction Generator

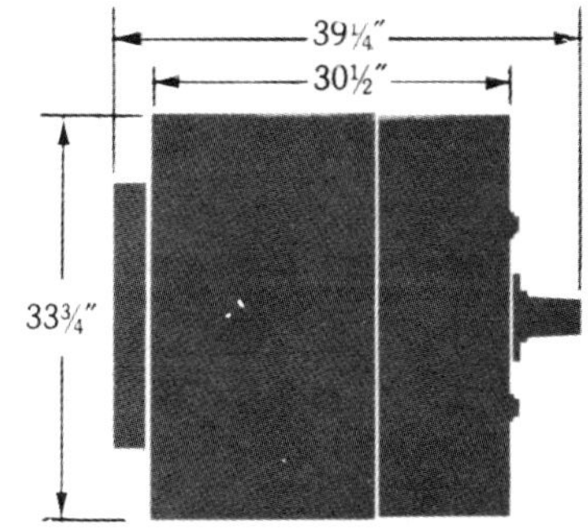

GT-603 Traction Generator

Dimensions shown are approximate. See outlines above for details.

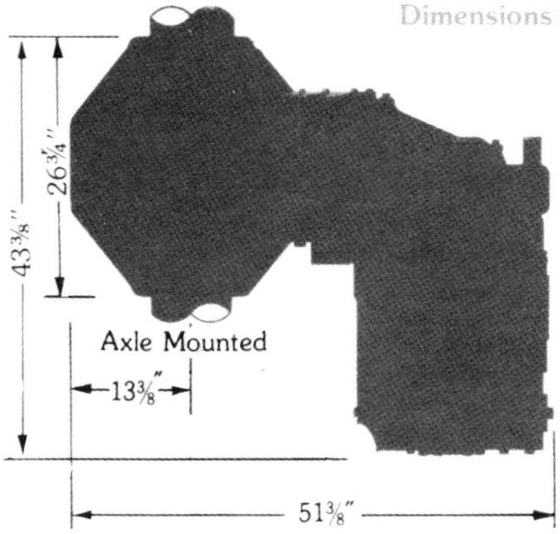

GE-763 Traction Motor And GA-70 Gear Unit

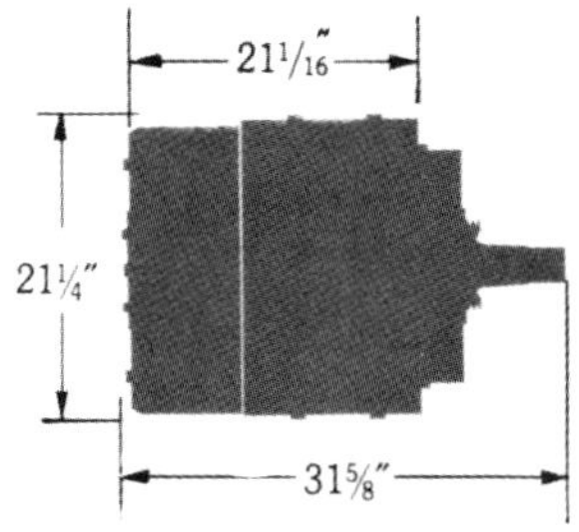

GE-763 Traction Motor

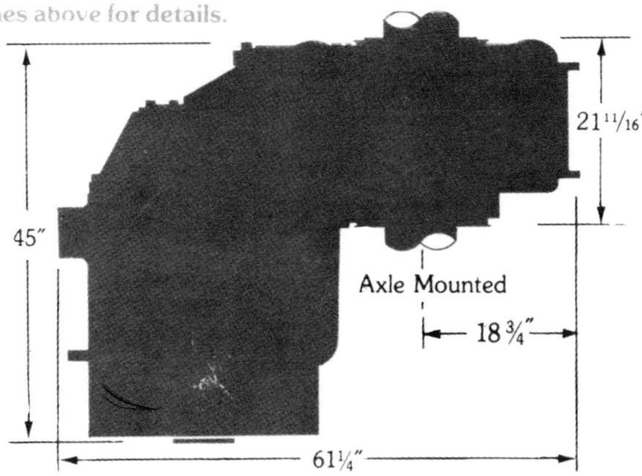

GE-761 Traction Motor and GA-72 Gear Unit

DIESEL-ELECTRIC EXPORT LOCOMOTIVES

Summary Model U30C AC/DC Diesel-Electric Locomotive

RATINGS	ENGLISH	METRIC (S.I.)
Diesel engine brake horsepower (useful service output under U.I.C. standard conditions)	3250 hp	2434 kw
Continuous horsepower to alternator for traction	3000 hp	2238 kw
Tractive effort at 30% adhesion (nominal weight)	79.200 lbs	35.925 kg
Continuous tractive effort (93:18 gear ratio)	59.790 lbs	27.120 kg
Maximum locomotive speed with new or worn wheels 36 inch (914 mm) (93:18 gear ratio)	64 mph	103 KM/H
Locomotive speed-tractive effort curve	(see next page)	

WEIGHTS		
Nominal locomotive (fully loaded)	210.000 lbs	95.256 kg
Per driving axle (fully loaded)	35.000 lbs	15.876 kg
Locomotive weight subject to manufacturing tolerance of ±2%. Modifications and gages above 42 inches (1067 mm) may increase weight.		

WHEEL ARRANGEMENT	C-C	

MAJOR DIMENSIONS		
Length over end frames	55 ft. 6 in	16.916 mm
Height over operator's cab	12 ft. 1-1/2 in	3.697 mm
Width over cab	8 ft. 11 in	2.718 mm
Width over platform	9 ft. 0 in	2.743 mm
Clearance under gear case (with 36-inch wheels)	4-1/4 in	108 mm
Locomotive outline drawing	(see pages 6-7)	
Minimum radius of curvature	186 ft	56.7 m

CAPACITIES		
Fuel	1200 U.S. gal	4542 liters
Lubricating oil	320 U.S. gal	1211 liters
Engine water	210 U.S. gal	795 liters
Sand	18 cu ft	510 liters

Introduction

DESIGN - The U30C is designed for service in freight, passenger and switching traffic. The universal design permits adaptability to railways worldwide; any track gage from 39-3/8 (1000 mm) to 66 inches (1676 mm). The operating cab, located between the power plant hood and front (low) equipment hood, provides visibility in either direction. The control station is at the right side with the front hood leading.

POWER - A diesel engine is the primary power source. A traction alternator is connected to the engine and furnishes power through rectifiers to the axle-mounted traction motors. The electric transmission permits high utilization of the horsepower output throughout the locomotive speed range. Speed is controlled by a throttle lever that regulates engine output and controls the proper application of power to the traction motors. The direction of motion is controlled by a reverse lever. The throttle and reverse levers are interlocked to prevent reversal under power.

MATERIALS - All materials are in accordance with standard material specifications of General Electric. High standards of quality control are maintained. Materials and specifications are subject to change without notice.

TESTING - All component parts are given standard commercial tests before assembly on the locomotive. Each complete locomotive is tested as follows:

1. Control wiring is checked by observing sequence of contactor and relay operation and testing continuity of circuit between terminals.
2. High-potential tests of traction and control circuits are made in accordance with current U.S.A. standards.
3. Air brake tests provide satisfactory operation of the system and check the piping.
4. The power plant is tested at full load to check and adjust alternator characteristics and engine performance, including power and speed.

PAINTING - Interior: gray. Underframe and running gear: black. Interior of battery compartment: special acid resisting paint. Exterior: finish painted in customer's color choice. A Color Selector is furnished to suggest painting schemes.

EXPORT SHIPMENT - For overseas delivery, superstructure is protected for on-deck ocean shipment. Running gear is packed for below-deck shipment.

Contents

Summary	4
Locomotive Performance	5
Optional Gearing	5
Superstructure	6
Location of Major Systems	6-7
Power Plant	8
Electric Transmission	9
Running Gear	9
Operating Controls	10
Locomotive Accessories	10
Locomotive Brakes	10
Modifications	10-11

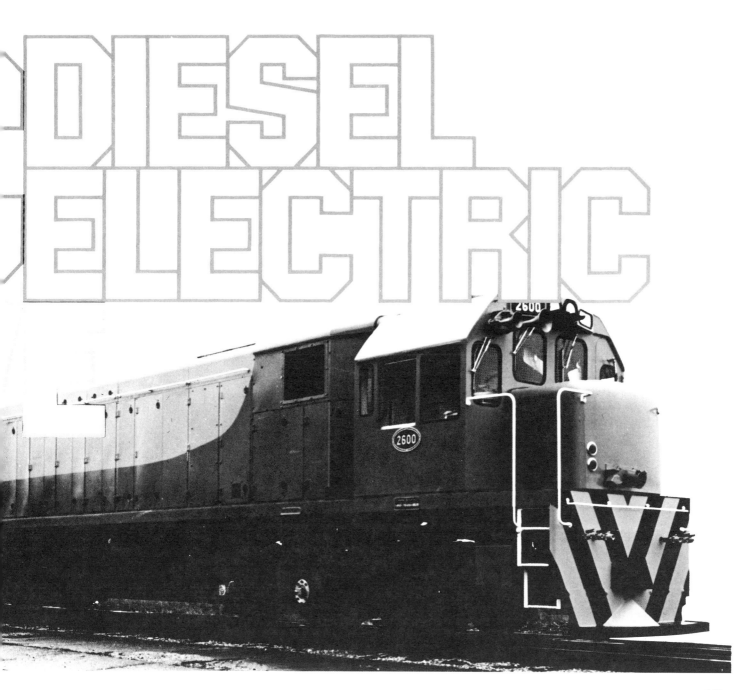

Superstructure

The superstructure, of welded steel construction, consists of a low front hood, an operator's cab, engine hood, and a radiator hood. All hoods are bolted to the underframe and are removable.

UNDERFRAME - Consists of two steel main sills with end plates, deck plates and transverse steel bolsters, securely welded in place. The center and side bearing plates are of wear-resistant steel.

FRONT HOOD - Contains air brake equipment and sand boxes. Doors provide access to this equipment.

OPERATOR'S CAB - Sides and roofs are insulated against heat and sound. The floor, raised above the platform, is covered with heavy-duty composition material. The cab has safety glass windows in the front, rear and each side for visibility in all directions. Center windows on each side have sliding sash, equipped with latches. All other windows are fixed and mounted in rubber self-sealing sash. One door at each end of the operator's cab provides access to walkways along the hoods. The doors have windows, weather stripping and provision for locking.

CONTROL COMPARTMENT - Located behind the operator's cab and encloses control devices and main power switching equipment.

BATTERY BOX - The batteries are located on the left side walkway, just behind the operator's cab.

ENGINE HOOD - Encloses the diesel engine, traction alternator set, and equipment blower. Side doors and roof hatches provide access to the equipment. Equipment cooling air is filtered by self-cleaning inertial air filters. Engine combustion air is taken from outside the hood and cleaned by inertial (primary) and paper (secondary) filters.

RADIATOR HOOD - Contains radiator, fan, air compressor, and sand boxes. Doors for access are provided.

WALKWAYS - Provided at each end of the locomotive and along the hoods. Sidesteps are located at each corner for boarding. Walkways and steps have handrails and non-skid treads.

PILOTS - Bolted to each end plate.

COUPLERS - AAR Type E top-operated couplers with rubber-cushioned draft gear are provided when center couplers are required.

LIFTING AND JACKING - Four combination jacking pads and lifting lugs are provided on the underframe.

FUEL TANK - Fabricated of heavy gage steel, well baffled, vented and bolted to the underframe. Provision is made for draining and cleaning. Filler connections and fuel level gages are furnished on each side of the locomotive.

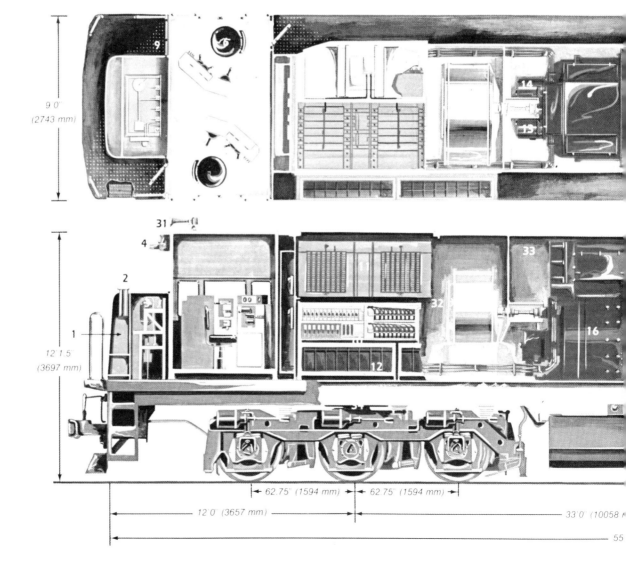

Maximum Equipment Diagram with 36 inch (914 mm) wheels

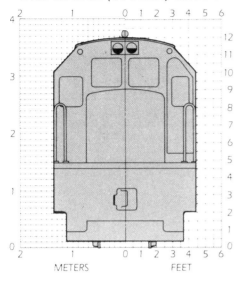

Location of Major Systems

1. Sand Box
2. Sand Box Fill
3. Air Brake Equipment
4. Headlight
5. Cab Heater (optional)
6. Cab Seat
7. Controller
8. Brake Stand
9. Hand Brake
10. Control Compartment
11. Dynamic Brake (optional)
12. Battery
13. Exciter
14. Battery Charging Generator
15. Equipment Blower
16. Traction Alternator
17. Engine
18. Fuel Tank
19. Turbocharger
20. Engine Water Tank
21. Engine Inlet Air Filters
22. Lubricating Oil Filter
23. Lubricating Oil Cooler
24. Air Compressor
25. Radiator Shutter
26. Radiator
27. Radiator Fan
28. Air Reservoir
29. Floating Bolster Truck
30. Traction Motor
31. Horn
32. Equipment Air Filters
33. Rectifiers
34. Lifting And Jacking Pads

Note: Second control stand is optional.

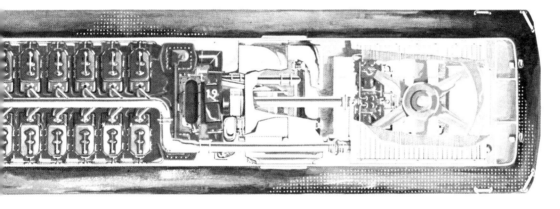

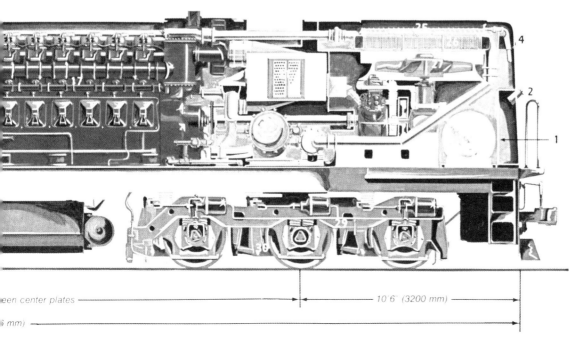

Performance Features

LOW FUEL RATE - Low auxiliary load with eddy current clutch drive to radiator fan assures that the fan only requires power when the radiator water must be cooled. Turbocharged, 4-stroke-cycle diesel engine provides highly-efficient operation.

HIGHER ADHESION - Floating bolster trucks with traction motors mounted in tandem minimize weight shift and wheel slippage. The GTA-11 alternator has adequate capacity so that all motors are connected in parallel across the alternator field at all times – from start to maximum speeds. The connection assures that, when one wheel slips, the power is inherently transferred to the remaining five motors equally. Sentry wheelslip detection and correction system furnishes the speed and accuracy of action to correct any wheelslip that occurs.

CLEAN AIR - Ambient air for the traction motors and alternator is first drawn through self-cleaning inertial air filters which purge the air of a high portion of impurities. Engine intake air is further filtered through paper filters, thus maintaining a supply of clean air to the diesel engine.

ELECTRICAL ADVANCES - Constant Horsepower Excitation Control (CHEC) is a modern static power train control system. It sets and controls horsepower at the maximum level consistent with the requirements of all subsystems. Flashovers are virtually eliminated since all traction motor operation is in the full field mode. Shunt field operation is not necessary due to the higher voltage capability of the GTA-11 CHEC Excitation System.

Power Plant

DIESEL ENGINE -

Type	One GE FDL-12
Brake horsepower	3300
Number of cylinders	12
Cylinder arrangement	45 V
Stroke cycle	4
Bore and stroke	9 inches (228.6 mm) x 10-1/2 inches (266.7 mm)
Full load speed	1050 rpm
Turbocharger	One

GOVERNOR - Self-contained, electro-hydraulic modulating engine governor automatically regulates horsepower output at each throttle setting. With low oil or water pressure, the governor reduces engine load to match available pressure until it returns to normal. With loss of oil or water pressure, the engine is shut down.

OVERSPEED PROTECTION - Engine is automatically shut down if speed exceeds maximum rated rpm by 10 percent.

ENGINE AIR FILTER - The engine air intake is equipped with self-cleaning inertial primary and secondary paper filters.

COOLING SYSTEM - Water is circulated through the engine, turbocharger, intercooler, radiator, and lube oil cooler by a gear-driven centrifugal pump integral with the engine. An expansion tank with sight gage indicates water level. The water fill is located on roof. Abnormally low water pressure automatically shuts down the engine.

ENGINE WATER TEMPERATURE - Automatically controlled.

FUEL SYSTEM - A motor-driven pump transfers fuel from the tank through filters to the injection pumps. Each cylinder is equipped with high-pressure fuel injection pump and injector.

LUBRICATING SYSTEM - A single full-flow pressure-regulated system is supplied by a gear-type pump integral with the engine. A reservoir is located in the engine sub-base. Filters and water-cooled oil cooler are provided. Abnormally low lube oil pressure automatically shuts down the engine.

ENGINE STARTING - Cranking windings are on the auxiliary generator and exciter.

HORSEPOWER RATING - Useful service output under U.I.C. (International Railway Union) standard conditions.

Electric Transmission

TRACTION MOTORS - Six GE-761 traction motors are furnished. They are direct current, series wound, separately ventilated. Armatures are mounted in anti-friction bearings. Motors drive through single-reduction spur gearing. They are supported by the axles to which they are geared and by resilient nose suspensions on truck transoms.

TRACTION ALTERNATOR - One GTA-11 traction alternator is engine mounted. It is an AC, single anti-friction bearing, separately excited machine. Output is rectified by a full-wave rectifier.

CONTROL - Railway-type single-station single-unit control is basic equipment. Control devices are grouped in dust and water tight steel compartments, fitted with access doors. Reverser and braking switch are electro-pneumatically operated. Contactors are magnetically operated. Circuit breaker-type switches are used in control circuits where overload protection is required. Transition and field shunting are not used.

EXCITER - One GY-27 is gear-driven from the traction alternator and provides controlled excitation of the alternator field.

BATTERY CHARGING GENERATOR - One Type GY-27 generator is gear-driven from the alternator and furnishes power at regulated potential for battery charging, lighting and control.

STORAGE BATTERY - A 32-cell lead-acid battery is furnished for starting the engine and supplying power for lights and other auxiliaries when the engine is shut down.

EQUIPMENT BLOWER - Direct-driven blower supplies ventilating air to the alternator, rectifier, auxiliary generator, exciter, and traction motors through platform ducting and flexible connections.

GROUND RELAY PROTECTION - If a ground occurs, engine speed returns to idle, power is removed and visual as well as audible indication is given.

SENTRY WHEELSLIP CORRECTION - Wheelslip is automatically detected by comparison of output signals from speed sensors located in each traction motor. Slip is corrected by automatic application of sand and reduction of power.

Running Gear

Consists of two, three-axle lateral motion swivel trucks. A tandem motor arrangement provides low weight transfer and excellent adhesion. Centerplate load is distributed by the "floating bolster" to four rubber mounts which rest on the truck frame and provide controlled lateral motion. Truck frame consists of cast-steel side frames joined integrally with structural steel shapes by electric welding. It is supported by alloy steel coil springs over the journal boxes. Friction-type snubbers damp vertical and lateral oscillation.

WHEELS - Solid multiple wear, rolled-steel of 36-inch (914 mm) diameter, 2-1/2 (63.5 mm) thick rims. The wheels have AAR standard tread and flange contour.

AXLES - Forged carbon steel to AAR material specifications.

JOURNALS - Equipped with sealed, grease-lubricated roller bearings. Guides are lined with renewable wear-resistant plates.

PEDESTAL GUIDES - Renewable wear-resistant liners are bolted to pedestal guides.

CENTER PLATES - Equipped with renewable wear-resistant liners and arranged for lubrication.

SIDE BEARINGS - Provided with renewable wear-resistant wear plates.

SAFETY BRACKETS - Prevent slewing and permit the trucks to be lifted with the superstructure.

Operating Controls

Controls and instruments are grouped at the operator's station and switch panel and gage panel in the operator's cab.

OPERATING CONTROLS:
Master controller with throttle, reverser, and dynamic braking selector levers
Engine control switch
Brake valves
Sander valve
Bell ringer valve
Air horn valve
Window wiper valves
Circuit breakers and switches
Emergency fuel shutoff
Emergency engine stop switch

INSTRUMENTS:
Brake gages
Fuel oil pressure gage
Engine intake manifold pressure gage
Lube oil pressure gage

WARNING INDICATORS:
Low engine lubrication oil pressure – alarm bell and green light
Low engine water pressure – alarm bell and yellow light
Crankcase overpressure – alarm bell and red light
High engine water temperature – alarm bell and red light
Wheelslip – buzzer and white light
Engine shutdown – alarm bell and no charge light
Ground relay – alarm bell and white light
Battery not charging – alarm bell and blue light
Alternator overload – alarm bell and yellow light
Rectifier overtemperature – alarm bell and red light
Power reduction – yellow light

Accessories

AIR FILTER (BRAKE SYSTEM) - Centrifugal, with replaceable element and automatic drain valve.
AIR FILTER (AUXILIARY AIR DEVICES) - Centrifugal, with replaceable element and automatic drain valve.
ARM RESTS - Two, window mounted.
AUTOMATIC DRAIN VALVES - One at each main reservoir.
BELL - One, stationary, with air-operated ringer and operating valve.
EMERGENCY BRAKE VALVE - At helper's station.
EMERGENCY FUEL SHUTOFF - Three, one on each side of the underframe and one on the engine control panel.
EXTENSION LAMP RECEPTACLES - Two, in control compartment and engine hood, with one lamp and 35-foot cable.
FIRE EXTINGUISHER - Two, five-pound dry chemical.
FUEL GAGES - One on each side of locomotive near fill pipe.
HEADLIGHTS - Electric, at each end of the locomotive. Each consists of two 200-watt, 30-volt, sealed-beam lamps. Dimming control is provided.
HORN - One, air-operated, single-tone.
INTERIOR LIGHTS - Electric, for operating cab, hoods and instruments.
MARKER LIGHTS - Four, red, single-aspect electric lights, two at each end of the locomotive.
SANDERS - Eight, pneumatically operated, arranged to sand ahead of the lead wheels in each direction.
SEATS - Two, swivel type, with back rests, adjustable for height and located to enable operation in either direction.
SUN VISORS - Two, adjustable-type.
WATER TEMPERATURE GAGE - Located in engine cab.
WINDOW WIPERS - Six, air-operated, mounted on front and rear windows of operating cab.

Locomotive Brakes

AIR BRAKES - Schedule 26L with 26F control valve combined independent and automatic is basic equipment. Compressed air locomotive brakes may be operated either independently or with train brakes. Connections for furnishing compressed air to the train brakes are provided at each end of the locomotive.
COMPRESSOR - One three-cylinder, two-stage, water cooled engine-driven air compressor furnishes air for the locomotive and train braking systems.
Compressed air displacement:
Idle engine speed .... 101 cfm
(2860 liters/min)
Full engine speed .... 236 cfm
(6690 liters/min)
RESERVOIRS - 40,000 cubic inch (655 liters) capacity for storing and cooling air for the brake system.
BRAKE EQUIPMENT - Brake cylinders are mounted on the running gear and operate equalized brake rigging, which applies braking to each driving wheel. Adjustment is provided to compensate for wheel and shoe wear. There is one brake shoe per wheel.
HAND BRAKE - Located in the operator's cab for holding the locomotive at standstill.

Modifications

ADDITIONAL FIRE EXTINGUISHERS - To meet requirements.
ADDITIONAL FUEL - Total capacity can be increased to 2000 gallons (7570 liters).
ADDITIONAL RESERVOIR CAPACITY - Total capacity can be increased to 50,000 cubic inches.

ADDITIONAL SEAT - A third seat in the operating cab.
AIR CONDITIONING - For operator's cab.
ALKALINE-TYPE BATTERY - Instead of the lead-acid battery.
BRAKE CHARGING AMMETER - One, on back wall of operator's cab.
BATTERY CHARGING RECEPTACLE - Can be mounted on one or both sides of the locomotive for battery charging from a wayside source.
CAB HEATERS WITH DEFROSTERS - One or two can be installed in the operating cab.
CAB ROOF VENTILATORS - Adjustable for either direction of operation.
CAB VENTILATING FANS - Mounted in operator's cab.
CAB WINDOW AWNINGS - On each side of operating cab if maximum equipment diagram permits installation.
COLOR CODING - Applied to pipe connections according to requirements.
COUPLER ARRANGEMENT - To meet requirements.
COUPLER HEIGHT - To meet customer requirements.
CREW LOCKER - In operator's cab.
DYNAMIC BRAKING - Equipment for braking electrically, using traction motors as generators and dissipating the electric power in forced-ventilated resistors. Interlocking prevents application of locomotive air brakes during dynamic braking. Dynamic braking is overridden during emergency brake applications.
EMERGENCY SANDING - In addition to manually operated valve, sanding can be automatically initiated by an emergency brake application.

FLANGE LUBRICATORS - Four or eight, attached to running gear. (Availability is dependent on truck configuration.)
HEAD-END POWER - Up to 15 kilowatts at 75 volts for caboose. Up to 400 kilowatts at 1500 volts for air conditioning and train heat.
HORN - Customer's choice instead of single-tone horn.
HOT PLATE - Electric, in operator's cab.
HUBODOMETER - Axle-drive, for distance recording.
ILLUMINATED NUMBER BOXES - Front and rear.
INCREASED LOCOMOTIVE WEIGHT - Heavier axle loadings for higher adhesive weight.
LEFT SIDE CONTROL STATION - Control station at left side of operating cab with front (low) hood leading, instead of right side. Controls can also be located on either side with long hood leading.
LOADMETER - To indicate order of magnitude of the tractive effort being maintained.
LOCOMOTIVE OVERSPEED PROTECTION - Returns engine to idle, automatic brake application.
LOW WATER LEVEL ALARM.
MARKER LIGHTS - Multicolor available, two or three color.
MOTOR CUT-OUT SWITCH - Any traction motor may be cut out individually.
MULTIPLE-UNIT CONTROL - To enable the operation of two or more locomotive units from one operator's station.
SAFETY AND/OR VIGILANCE CONTROL - Foot-suppression or other safety and/or vigilance controls to give a service train-brake application and return engine to idle, after short warning period.

SPEEDOMETER - Electric, with provision to compensate for wheel wear.
SPEED RECORDER - One combination speedometer, speed recorder and odometer.
SUN VISORS - Additional, fully adjustable sun visors for the operator's cab.
SWITCHMEN'S END STEPS - At each end of the locomotive.
TOOL BOX - Containing emergency hand tools.
TWO-STATION CONTROL - Two control stations for operating from either of two diagonally opposite positions.
WATER-COOLED AIR COMPRESSOR - Three cylinder available.
WATER COOLER - Either an electric cooler or an insulated water tank.
WAYSIDE LIGHTING - Receptacle on one or both sides of locomotive for lighting from wayside a-c source.
WHEELS - Diameter—to meet requirements, from 36 inches (914 mm) to a maximum of 40 inches (1016 mm). Type—Steel-tired wheels with locking rings or extra thick rims for later application of steel tires by the customer.

Vacuum Brakes:
(A) Vacuum or Compressed Air Train Brakes—Schedule 28L-AV1 independent and automatic locomotive compressed air brakes. Six-cylinder compressor-exhausters are recommended for locomotives which will also haul compressed air brake trains.
(B) Vacuum Train Brakes—Schedule 28L-V1 independent and automatic locomotive compressed air brakes and vacuum train brakes.
Alternate Air Brake - Schedule 26LA instead of 26L.

Compressor-Exhausters Available	Idle/Full Engine Speed	Idle/Full Engine Speed
6-Cylinder	Cfm	Liters per Min
Compressed Air (two cylinders)	69/161	1950/4500
Vacuum (four cylinders)	276/644	7820/18240

Locomotive Performance

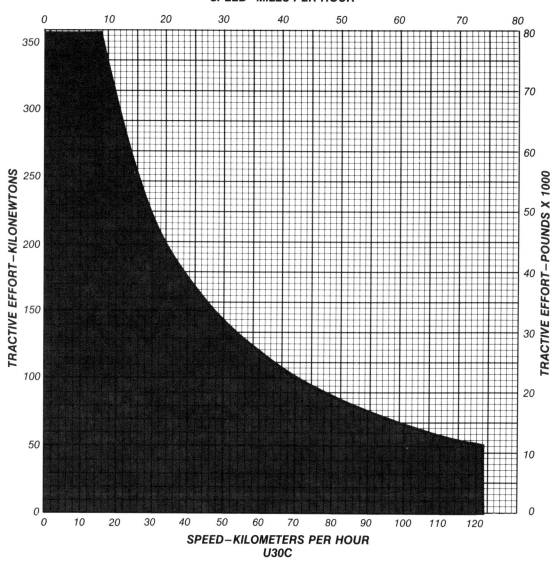

U30C

Optional Gearing
WITH GE-761 MOTORS

Wheel Diameter Inches	Gear Ratio	Maximum Locomotive Speed With New Or Worn Wheels		Tractive Effort At Continuous Motor Rating		
		MPH	KM/H	LB	KN	Kg
36	93:18	64	103	59,790	265	27,120
36	92:19	68	109	56,060	249	25,429
38	93:18	67	108	56,650	252	25,696
38	92:19	72	116	53,080	235	24,077
40	93:18	71	114	53,810	239	24,408
40	92:19	76	122	50,440	224	22,979

Note: Alternate gear ratios are available for special applications.

Summary Model U11B Diesel-Electric Locomotive

RATINGS	ENGLISH	METRIC (S.I.)
Diesel engine brake horsepower (useful service output under U.I.C. standard conditions - 4 hour rating)	1100 hp	821 kw
Continuous horsepower to generator for traction	1000 hp	746 kw
Tractive effort at 30% adhesion (minimum weight)	37,200 lb	165 kn
Continuous tractive effort	36,200 lb	161 kn
Maximum locomotive speed with new or worn wheels 36 inch (914 mm) (93:18 gear ratio)	64 mph	103 kph
Locomotive speed tractive effort curve	(see page 5)	

WEIGHTS

Minimum locomotive (fully loaded)	124,000 lb	56,246 kg
Per driving axle (fully loaded)	31,000 lb	14,061 kg

Locomotive weight subject to manufacturing tolerance of +2%. Modifications and gages above 42 inches (1067 mm) may increase weight.

WHEEL ARRANGEMENT .... B-B

MAJOR DIMENSIONS

Length over end frames	38 ft, 4 in	11,684 mm
Height over operator's cab	12 ft, 0 in	3658 mm
Width over cab	9 ft, 0 in	2743 mm
Width over platform	9 ft, 0 in	2743 mm
Clearance under gear case (with 36-inch wheels)	4 1/4 in	108 mm
Locomotive outline drawing	(see page 6)	
Minimum radius of curvature locomotive alone	75 ft	22.9 m

CAPACITIES

Fuel	400 U.S. gal	1514 liters
Lubricating oil	120 U.S. gal	454 liters
Engine water	140 U.S. gal	530 liters
Sand	12 cu ft	340 liters

Introduction

DESIGN - The U11B is designed for service in freight, passenger and switching traffic. The universal design permits adaptability to railways worldwide; any track gage from 36 inches (914 mm) to 66 inches (1676 mm). The operating cab located ahead of the power-plant hood provides visibility in either direction. The control station is at the right side with the operating cab leading.

POWER - A diesel engine is the primary power source. A traction generator is the primary power source. A traction generator is connected to the engine and furnishes power to the axle-mounted traction motors. The electric transmission permits high utilization of the horsepower output throughout the locomotive speed range. Speed is controlled by a throttle lever that regulates engine output and controls the proper application of power to the traction motors. The direction of motion is controlled by a reverse lever. The throttle and reverse levers are interlocked to prevent reversal under power.

MATERIALS - All materials are in accordance with standard material specifications of General Electric. High standards of quality control are maintained. Materials and specifications are subject to change without notice.

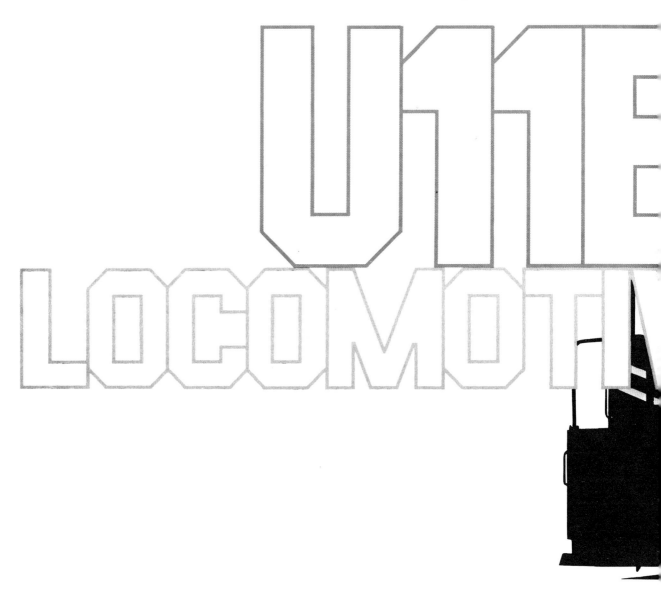

TESTING - All component parts are given standard commercial tests before assembly on the locomotive. Each complete locomotive is tested as follows:

1. Control wiring is checked by observing sequence of contactor and relay operation and testing continuity of circuit between terminals.
2. High-potential tests of traction and control circuits are made in accordance with current U.S.A. standards.
3. Air brake tests provide statisfactory operation of the system and check the piping.
4. The power plant is tested at full load to check and adjust generator characteristics and engine performances, including power and speed.

PAINTING - Interior: gray. Under frame and running gear: black. Interior of battery compartment: special acid resisting paint. Exterior: finish painted in customer's color choice. A Color Selector is furnished to suggest painting schemes.

EXPORT SHIPMENT - For overseas delivery, superstructure is protected for on-deck ocean shipment. Running gear is packed for below-deck shipment.

Contents

Summary 4
Locomotive Performance 5
Optional Gearing 5
Superstructure 6
Location of Major Systems 6-7
Power Plant 8
Electric Transmission 8
Running Gear 9
Operating Controls 10
Locomotive Accessories 10
Locomotive Brakes 10
Modifications 11

DIESEL ELECTRIC

Superstructure

The superstructure, of welded steel construction, consists of an operator's cab and engine hood. The hood is bolted to the underframe and removable.

UNDERFRAME - Consists of two steel main sills with end plates, deck plates and transverse steel bolsters, securely welded in place. The renewable side bearing plates are of wear-resistant steel. The center plates are equipped with renewable wear-resistant steel liners.

OPERATOR'S CAB - Sides and roof of the operator's cab are insulated against heat and sound. The floor, raised above the platform, is covered with heavy-duty composition material. The cab has safety glass windows in the front, rear and each side for visibility in all directions. Center windows on each side of the cab have sliding sash, equipped with latches. All other windows are fixed and mounted in rubber self-sealing sash. One door at each end of the operator's cab provides access to walkways along the hoods. The doors have windows, weather stripping and provision for locking.

ENGINE HOOD - Encloses the diesel engine and traction generator set, a traction motor blower air compressor, air brake equipment, radiator, radiator fan, and air reservoirs. Louvers with panel-type filters in the side doors provide ventilating air for the rotating equipment. Engine air is cleaned by inertial pre-cleaners and dry type filters.

WALKWAYS - Provided at each end of the locomotive and along the hoods. Sidesteps are located at each corner for boarding. Walkways and steps have handrails and non-skid treds.

PILOTS - Bolted to each end plate.

COUPLERS - AAR Type E top-operated couplers with rubber-cushioned draft gear are provided when center couplers are required.

LIFTING AND JACKING - Four combination jacking pads and lifting lugs are provided on the underframe (at the edge of the platform at the centerline of each bolster).

FUEL TANK - Fabricated of heavy gage steel, well baffled, vented, and bolted to the underframe. Provision is made for draining and cleaning. Filler connections and fuel level gages are furnished on each side of the locomotive.

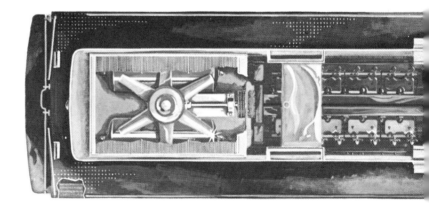

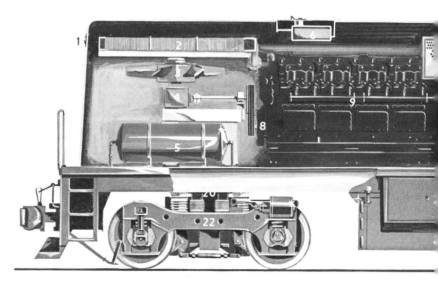

Maximum Equipment Diagram with 36 inch (914 mm) wheels

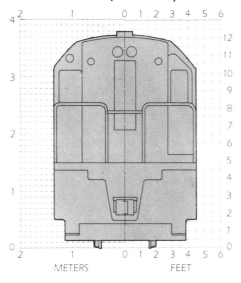

Location of Major Systems

1. Headlight
2. Radiator
3. Radiator Fan
4. Radiator Fan Gear Box
5. Main Air Reservoirs
6. Expansion Tank
7. Turbochargers
8. Engine Air Intake Filter
9. Diesel Engine
10. Traction Generator
11. Exciter-Battery Charging Generator
12. Dynamic Brake (if used)
13. Air Compressor
14. Air Brake Equipment
15. Traction Motor Blower
16. Control Compartment
17. Controller
18. Cab Seat
19. Handbrake
20. Jacking Pad and Lifting Lug
21. Traction Motor
22. Floating Bolster Truck
23. Batteries
24. Fuel Tank
25. Gage Panel
26. Brake Valves
27. Sand Box Fill

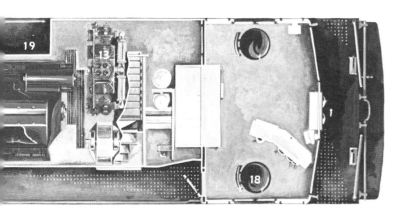

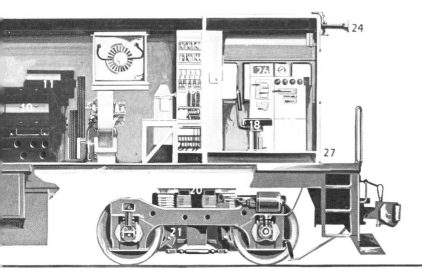

Performance Features

SUPERIOR RIDING QUALITIES - The rubber-mounted lateral motion trucks (bogies) offer superior riding qualities. They also enable the locomotive to easily follow track with many curves and uneven areas.

SIMPLICITY AND MAINTAINABILITY - Electrical systems are easily maintained by all levels of railroad electricians. The equipment arrangement is designed so that all equipment is readily accessible for maintenance.

Power Plant

DIESEL ENGINE -
Type . . . . One Caterpillar D399
Brake horsepower
  (Four-hour rating) 1100
Number of cylinders 16
Cylinder arrangement . . . . 60°V
Stroke cycle 4
Bore and stroke . 6 1/4 inches (159 mm)
            x 8 inches (203 mm)
Full load speed . . . . 1300 rpm
Turbochargers . . . . Two
Aftercoolers . . . . Two

GOVERNOR - Self-contained, hydro-mechanical, spring-balanced, flyball governor maintains engine speed at each throttle setting.

OVERSPEED PROTECTION - Engine is automatically shut down if speed exceeds maximum rated rpm by 10%.

ENGINE AIR FILTER - The engine air intake is equipped with inertial pre-cleaners and dry type filters.

COOLING SYSTEM - Water is circulated through the engine radiator, aftercooler, and lube oil cooler by a gear-driven centrifugal pump integral with the engine radiator cooling air is supplied by an engine-driven fan. An expansion tank with sight gage indicates water level. The water-fill is located on the locomotive roof.

ENGINE WATER TEMPERATURE CONTROL - Automatically controlled.

FUEL SYSTEM - A gear-type, engine-driven pump transfers fuel from the tank through filters to the injection pumps. An individual pump for each cylinder supplies fuel under high pressure to the injector.

LUBRICATING SYSTEM - A single, pressure-regulated system is supplied by a gear-type pump integral with the engine. A reservoir is located in the engine sub-base. Filters and water-cooled oil cooler are provided. Abnormally low lube oil pressure automatically shuts down the engine.

ENGINE STARTING - Engine is cranked by the traction generator from storage battery power.

HORSEPOWER RATING - Useful service output under U.I.C. (International Railway Union) standard conditions.

Electric Transmission

TRACTION MOTORS - Four GE-761 traction motors are furnished. They are direct current, series wound, separately ventilated. Armatures are mounted in anti-friction bearings. Motors drive through single-reduction spur gearing. They are supported by the axle to which they are geared and by resilient nose suspensions on truck bolsters.

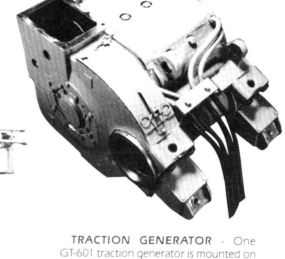

TRACTION GENERATOR - One GT-601 traction generator is mounted on an extension of the engine subbase and connected to it by means of a flexible coupling. The generator is a direct-current, two anti-friction bearing, self-ventilated, separately excited machine and is equipped with windings to permit starting the engine by storage battery power.

CONTROL - Railway-type single-station single-unit control is basic equipment. Control devices are grouped in dust and moisture-resistant steel compartments, fitted with access doors. Reverser and line contactors are electro-pneumatically operated. Other contac tors are magnetically operated. Circuit breaker-type switches are used in control circuits where overload protection is required. Transition is automatic.

EXCITER-BATTERY CHARGING GENERATOR - One GMG-175 combination exciter-battery charging generator set is belt-driven from the traction generator. The exciter provides controlled excitation of the traction generator field. The generator furnishes power at regulated potential for battery charging, lighting, and control.

STORAGE BATTERY - A 32-cell lead-acid battery is furnished for starting the engine and supplying power for lights and other auxiliaries when the engine is shut down.

TRACTION MOTOR BLOWER - A belt-driven traction motor blower supplies ventilating air to the traction motors through platform ducting and flexible connections.

AUTOMATIC WHEELSLIP DETECTION - Wheelslip is automatically detected with visual and audible warning given to the operator.

GROUND RELAY PROTECTION - If a ground occurs, power is removed and visual as well as audible indication is given to the operator.

Running Gear

Consists of two, two-axle lateral motion swivel trucks. Centerplate load is distributed by the cast-steel "floating bolster" to four rubber mounts which rest on the truck frame and provide controlled lateral motion. Truck frame consists of cast-steel side frames joined integrally with structural steel shapes by electric welding. It is supported by alloy steel coil springs over the journal boxes. Friction-type snubbers damp vertical and lateral oscillation.

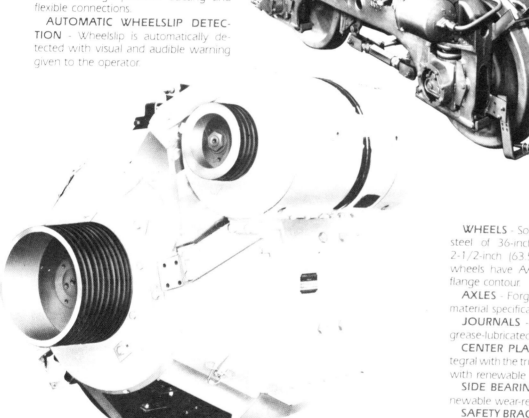

WHEELS - Solid multiple wear, rolled-steel of 36-inch (914 mm) diameter, 2-1/2-inch (63.5 mm) thick rims. The wheels have AAR standard tread and flange contour.

AXLES - Forged carbon steel to AAR material specifications.

JOURNALS - Equipped with sealed, grease-lubricated roller bearings.

CENTER PLATES - Center plates integral with the truck bolster are equipped with renewable wear-resistant liners.

SIDE BEARINGS - Provided with renewable wear-resistant wear plates.

SAFETY BRACKETS - Prevent slewing and to permit the trucks to be lifted with the superstructure.

Operating Controls

Controls and instruments are grouped at the operator's station and switch panel and gage panel in the operator's cab.

OPERATING CONTROLS:
Master controller with throttle, reversing lever, and dynamic braking selector lever
  Engine start push-button
  Brake valve
  Sander valve
  Bell ringer valve
  Air horn valve
  Window wiper valves
  Circuit breakers and switches
  Emergency fuel shutoff
  Engine stop switch

INSTRUMENTS:
  Brake gages
  Engine lubricating oil pressure gage
  Water temperature gage

WARNING INDICATORS:
  Low engine lubricating oil pressure — alarm bell and green light
  High engine water temperature — alarm bell and red light
  Wheelslip — buzzer and white light
  Ground relay — alarm bell and indicator
  Engine shutdown — alarm bell and no charge light
  Battery not charging — alarm bell and blue light

Accessories

AIR FILTER (BRAKE SYSTEM) - Centrifugal, with replaceable element and automatic drain valve.

AIR FILTER (AUXILIARY AIR DEVICES) - Centrifugal, with replaceable element and automatic drain valve.

EMERGENCY FUEL SHUTOFF - Three, one on each side of the underframe and one in the operator's cab.

EXTENSION LAMP RECEPTACLES - Two, in control compartment and engine hood, with one lamp and 35-foot cable.

FIRE EXTINGUISHER - One, five-pound dry chemical.

FUEL SIGHT GAGES - Two, one on each side of locomotive near fill pipe.

FUEL PRESSURE GAGE - One located on the engine.

HEADLIGHTS - Electric, at each end of the locomotive. Each consists of two, 200-watt, 30-volt, sealed-beam lamps. Dimming control is provided.

HORN - One, air-operated, single-tone.

INTERIOR LIGHTS - Electric, for operating cab, hoods and instruments.

MARKER LIGHTS - Four, red, single-aspect electric lights, two at each end of the locomotive.

SANDERS - Four, pneumatically operated, arranged to sand ahead of the lead wheels in each direction.

SEATS - Two, upholstered, swivel-type, with back rests, adjustable for height and located to enable operation in either direction.

SUN VISORS - Two, adjustable-type.

WINDOW WIPERS - Four, air-operated, mounted on windows for viewing in either direction of motion.

Locomotive Brakes

AIR BRAKES - Schedule 26NL combined independent and automatic is furnished as basic equipment. Compressed air locomotive brakes may be operated either independently or with train brakes. Connections for furnishing compressed air to the train brakes are provided at each end of the locomotive.

COMPRESSOR - One two-cylinder, two-stage, air cooled engine-driven air compressor furnishes air for the locomotive and train braking systems.
Compressed air displacement:
Idle engine speed . . . . 71 cfm
(2010 liters/min)
Full engine speed . . . . 142 cfm
(4020 liters/min)

RESERVOIRS - 40,000 cubic inch (655 liters) for storing and cooling air for the brake system.

BRAKE EQUIPMENT - Brake cylinders are mounted on the running gear and operate fully equalized brake rigging, which applies braking to each driving wheel. Adjustment is provided to compensate for wheel and shoe wear. There is one brake shoe per wheel.

HAND BRAKE - Located in the operator's cab for holding the locomotive at standstill.

Modifications

ADDITIONAL FIRE EXTINGUISHERS - To meet requirements.

ADDITIONAL FUEL - Additional capacity up to 600 gallons (2271 liters).

AIR CONDITIONING - For operator's cab.

ALKALINE-TYPE BATTERY - Instead of the lead-acid battery.

AUTOMATIC SANDING - In addition to manually operated valve, sanding can be automatically initiated by an emergency brake application.

BATTERY CHARGING AMMETER - One, on back wall of operator's cab.

BATTERY-CHARGING RECEPTACLE - Can be mounted on one or both sides of the battery compartment to provide battery charging from a wayside source.

CAB HEATERS WITH DEFROSTERS - Can be installed in the operating cab.

CAB ROOF VENTILATORS - Adjustable for either direction of operation.

CAB VENTILATING FANS - Mounted in operator's cab.

CAB WINDOW AWNINGS - On each side of operating cab if maximum equipment diagram permits installation.

CLASP BRAKES - Availability is dependent on specific truck configuration.

COLOR CODING - Applied to pipe connections according to requirements.

COUPLER ARRANGEMENT - To meet customer's requirements.

COUPLER HEIGHT - To meet customer requirements.

DYNAMIC BRAKING - Equipment for braking electrically, using the traction motors as generators and dissipating the electric power in forced-ventilated resistors. Interlock prevents application of automatic locomotive air brakes during dynamic braking. Dynamic braking is overridden during emergency brake applications.

EIGHT SANDERS - Four, additional, to provide sanding ahead of leading wheels of each truck. Additional sand capacity is provided.

FLANGE LUBRICATORS - Four or eight, attached to running gear. (Availability is dependent on truck configuration).

HEAD-END POWER - To suit requirements, within available space and weight limitations.

HORNS - Customer's choice instead of single-tone horn.

HOT PLATE - Electric, in operator's cab.

HUBODOMETER - Axle-driven, for distance recording.

INCREASED LOCOMOTIVE WEIGHT - Heavier axle loadings for higher adhesive weight.

LEFT SIDE CONTROL STATION - Control station at the left side of operating cab with this cab leading. Controls can also be located on either side with the engine hood leading.

LOADMETER - To indicate order of magnitude of the tractive effort being maintained.

LOCOMOTIVE OVERSPEED PROTECTION - Returns engine to idle, automatic brake application.

MOTOR CUT-OUT SWITCH - Motors in either truck may be cut out.

MULTIPLE-UNIT CONTROL - To enable the operation of two or more locomotive units from one operator's station.

SAFETY AND/OR VIGILANCE CONTROL - Foot-suppression or other safety and/or vigilance controls to give service train-brake application and remove power, after a short warning period.

SPEEDOMETER - Electric, with provision to compensate for wheel wear.

SPEED RECORDER - Combination speedometer, speed recorder and odometer.

SUN VISORS - Additional, full adjustable sun visors for the operator's cab.

TOOL BOX - Containing emergency hand tools.

TRACK GAGE (1000 MM AND ABOVE) - Locomotive can be furnished for any track gage from 39-3/8 inches (1000 mm) to a maximum of 66 inches (1676 mm).

TRACK GAGE (BELOW 1000 MM) - Locomotive can be furnished for any track gage below 39-3/8 inches (1000 mm) to a minimum of 36 inches (914 mm).

The GE-764 traction motor will be used for these gages. Continuous tractive effort will be 26,400 pounds (117 kN) with 63 mph (101 kph) gearing on 36-inch (914 mm) diameter wheels and 90:17 gear ratio.

TRUCKS - For applications with lower speed requirements rigid bolster side-equalized swivel trucks can be provided. The truck frame consists of structural steel shapes and plates joined integrally by electric welding. It is supported by alloy steel springs seated on drop-type equalizer bars. Friction-type snubbers damp vertical oscillation.

TWO-STATION CONTROL - Two control stations for operating from either one of two diagonally opposite positions.

WATER COOLER - Either an electric cooler or an insulated water tank.

WAYSIDE LIGHTING - Receptacle on one or both sides of battery compartment for lighting from wayside source.

WHEELS - Diameter - To meet the requirements, from 36 inches (914 mm) to a maximum of 40 inches (1016 mm). Type - Steel-tired wheels with locking rings or extra thick rims for later application of steel tires by the customer.

WHEELSLIP DETECTION - (Adhesion Loss Detection) — Instead of the field voltage drop comparison system, wheelslip may be automatically detected by comparison of output signals from an alternator mounted on each axle.

Alternate Brake Systems Available

Air Brakes
26LA or 26L

Vacuum Brakes:

(A) Vacuum or Compressed Air Train Brakes — Schedule 28L-AV1 independent and automatic locomotive compressed air brakes and vacuum or compressed air train brakes. The reservoir capacity is 40,000 cu. in. (654 liters) for air. Four-cylinder compressor-exhausters are recommended for locomotives which will also haul compressed air brake trains.

(B) Vacuum Train Brakes — Schedule 28L-V1 independent and automatic locomotive compressed air brakes and vacuum train brakes. The reservoir capacity is 40,000 cu. in. (654 liters) for air.

Compressor or Compressor-Exhausters Available	Idle/Full Engine Speed	Idle/Full Engine Speed
	Cfm	Liters per Min
3-Cylinder		
Compressord Air	111/222	3143/6286
3-Cylinder		
Compressed Air (one cylinder)	28/56	792/1585
Vacuum (two cylinders)	112/225	3170/6370
4-Cylinder		
Compressed Air (two cylinders)	76/152	2152/4304
Vacuum (two cylinders)	152/304	4304/8608

Universal Diesel-Electric Locomotives

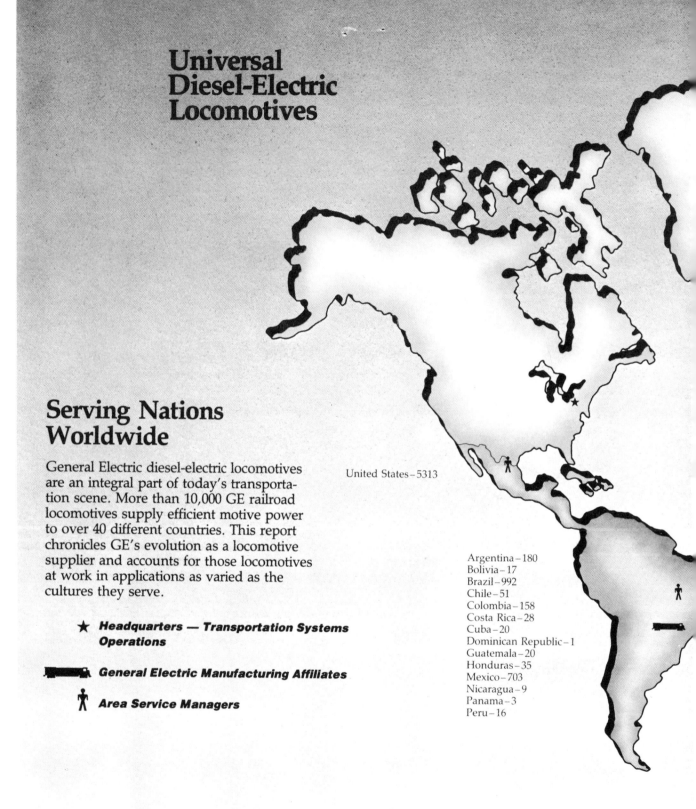

Serving Nations Worldwide

General Electric diesel-electric locomotives are an integral part of today's transportation scene. More than 10,000 GE railroad locomotives supply efficient motive power to over 40 different countries. This report chronicles GE's evolution as a locomotive supplier and accounts for those locomotives at work in applications as varied as the cultures they serve.

★ **Headquarters — Transportation Systems Operations**

General Electric Manufacturing Affiliates

Area Service Managers

United States – 5313

Argentina – 180
Bolivia – 17
Brazil – 992
Chile – 51
Colombia – 158
Costa Rica – 28
Cuba – 20
Dominican Republic – 1
Guatemala – 20
Honduras – 35
Mexico – 703
Nicaragua – 9
Panama – 3
Peru – 16

WORLD USERS

Algeria – 25
Angola – 69
Botswana – 14
Gabon – 40
Greece – 13
Ivory Coast – 1
Jordan – 36
Kenya – 36
Mozambique – 124
Nigeria – 51
South Africa – 1037

Spain – 76
Sudan – 50
Syria – 30
Tanzania – 14
Tunisia – 87
Turkey – 45
Zaire – 79
Zambia – 99
Zimbabwe – 10

Australia – 22
Bangladesh – 10
China – 420
Indonesia – 131
Japan – 1
Malaysia – 3
New Zealand – 55
Pakistan – 70
Philippines – 99
Thailand – 50
Vietnam – 48

Field Service Network: Backed by headquarters service and engineering specialists at GE's locomotive plant in Erie, Pennsylvania, General Electric service engineers provide technical support, advisory service, and any immediate assistance that might be required worldwide.

WORLD USERS

UNITED STATES

	QUANTITY	MODEL	GROSS HP	GROSS KW	GAUGE	YEAR
☐ Atchison, Topeka, & Santa Fe Ry. Co., The	3	B39-8	4000	2984	1.435m	1984
	16	B36-7	3940	2939	1.435m	1980
	69	B23-7	2490	1858	1.435m	1978-85
	157	C30-7	3280	2447	1.435m	1978-83
	100	U36C	3940	2939	1.435m	1972-75
	49	U23B	2490	1858	1.435m	1970-71
	25	U33C	3600	2886	1.435m	1969
	20	U23C	2490	1858	1.435m	1969
	6	U30CG	3280	2447	1.435m	1967-68
	10	U28CG	3050	2275	1.435m	1966
	16	U25B	2750	2052	1.435m	1962-63
☐ Auto Train	13	U36B	3940	2939	1.435m	1971-74
☐ Burlington Northern R.R. Co.	3	B32-8	3250	2425	1.435m	1984
	120	B30-7A	3210	2394	1.435m	1982-83
	8	B30-7	3280	2447	1.435m	1977-78
	242	C30-7	3280	2477	1.435m	1976-81
	9	U23C	2490	1858	1.435m	1969
	184	U30C	3280	2447	1.435m	1968-75
	66	U33C	3600	2686	1.435m	1968-71
	46	U30B	3280	2447	1.435m	1966-75
	16	U28B	3050	2275	1.435m	1966-67
	28	U28C	3050	2275	1.435m	1966
	42	U25C	2750	2052	1.435m	1964-65
	62	U25B	2750	2052	1.435m	1961-66
☐ Chessie System Railroads, The	64	B30-7	3280	2447	1.435m	1978-81
	35	U30B	3280	2447	1.435m	1971-75
	30	U23B	2490	1858	1.435m	1969
	13	U30C	3280	2447	1.435m	1967-68
	38	U25B	2750	2052	1.435m	1963-64
☐ Chicago, Milwaukee, St. Paul, Pacific	8	U30C	3280	2447	1.435m	1974
	5	U23B	2490	1858	1.435m	1973
	4	U36C	3940	2939	1.435m	1972
	4	U33C	3600	2686	1.435m	1968
	10	U30B	3280	2447	1.435m	1966-68
	12	U28B	3050	2275	1.435m	1966
	12	U25B	2750	2052	1.435m	1965
☐ Chicago & N.Western Transportation Co.	7	U30C	3280	2447	1.435m	1963
☐ Chicago, Rock Island, & Pacific	18	U30C	3280	2447	1.435m	1973
	25	U33B	3600	2686	1.435m	1968-69
	42	U28B	3050	2275	1.435m	1966-67
	39	U25B	2750	2052	1.435m	1963-65
☐ Consolidated Rail Corporation	22	C39-8	4000	2984	1.435m	1986
	25	C36-7	3940	2939	1.435m	1985
	10	C32-8	3250	2425	1.435m	1984
	50	C30-7A	3250	2425	1.435m	1984
	60	B36-7	3940	2939	1.435m	1983
	141	B23-7	2490	1858	1.435m	1977-79
	10	C30-7	3280	2447	1.435m	1977
	4	U36B	3940	2939	1.435m	1976
	99	U23B	2490	1858	1.435m	1972-77
	13	U36C	3940	2939	1.435m	1972
	33	U34CH	3760	2805	1.435m	1970-78*
	19	U23C	2490	1858	1.435m	1970
	81	U33B	3600	2686	1.435m	1968-70
	39	U33C	3600	2686	1.435m	1968-69
	10	U30C	3280	2447	1.435m	1967
	60	U30B	3280	2447	1.435m	1966-67
	15	U28C	3050	2275	1.435m	1966
	2	U28B	3050	2275	1.435m	1966
	20	U25C	2750	2052	1.435m	1964-65
	182	U25B	2750	2052	1.435m	1962-65
☐ Department of Transportation	1	U30C	3280	2447	1.435m	1971
☐ Detroit Edison Co.	11	U30C	3280	2447	1.435m	1970-75
☐ Gilford Transportation Industries, Inc.	10	U18B	2000	1492	1.435m	1975
	9	U33C	3600	2686	1.435m	1970
	16	U23B	2490	1858	1.435m	1968
	12	U30C	3280	2447	1.435m	1967-68

UNITED STATES

	QUANTITY	MODEL	GROSS HP	GROSS KW	GAUGE	YEAR
☐ Illinois Central Gulf Railroad Co.	10	U33C	3600	2686	1.435m	1968
	6	U30B	3280	2447	1.435m	1967
☐ Kaiser Steel Corp.	5	U30C	3280	2447	1.435m	1968
☐ Lake Superior & Ishpeming R.R. Co.	5	U23C	2490	1858	1.435m	1968-70
	2	U25C	2750	2052	1.435m	1964
☐ Missouri — Kansas — Texas R.R. Co.	3	U23B	2490	1858	1.435m	1973
☐ National Railroad Passenger Service	25	P30CH	3280	2447	1.435m	1975-76
☐ Norfolk Southern Corporation	139	C39-8	4000	2984	1.435m	1984-87
	22	B30-7A	3210	2394	1.435m	1982
	43	C36-7	3940	2939	1.435m	1981-84
	6	B36-7	3940	2939	1.435m	1981
	54	B23-7	2490	1858	1.435m	1978-81
	80	C30-7	3280	2447	1.435m	1978-79
	70	U23B	2490	1858	1.435m	1972-77
	10	U33C	3600	2686	1.435m	1970-72
	8	U30C	3280	2447	1.435m	1967-74
	110	U30B	3280	2447	1.435m	1967-71
	30	U28B	3050	2275	1.435m	1966
	16	U25B	2750	2050	1.435m	1962-65
☐ Pittsburgh & Lake Erie R.R. Co., The	22	U28B	3050	2275	1.435m	1966
☐ Providence & Worcester R.R. Co.	1	B23-7	2490	1858	1.435m	1978
	1	U18B	2000	1492	1.435m	1976
☐ Seaboard System Railroad, Inc.	120	B36-7	3940	2939	1.435m	1985
	17	B30-7	3280	2447	1.435m	1980
	95	C30-7	3280	2447	1.435m	1979-81
	55	B23-7	2490	1858	1.435m	1978-79
	105	U18B	2000	1492	1.435m	1973-74
	7	U36C	3940	2939	1.435m	1971
	98	U23B	2490	1858	1.435m	1970-75
	108	U36B	3940	2939	1.435m	1970-72
	29	U33B	3600	2686	1.435m	1967-69
	81	U30C	3280	2447	1.435m	1966-72
	29	U30B	3280	2447	1.435m	1966-67
	12	U28C	3050	2275	1.435m	1966
	47	U25C	2750	2052	1.435m	1963-65
	27	U25B	2750	2052	1.435m	1963-64
☐ Soo Line R.R. Co.	10	U30C	3280	2447	1.435m	1968
☐ Southern Pacific Transportation Co.	20	B36-7	3940	2939	1.435m	1980-84
	15	B23-7	2490	1858	1.435m	1980
	108	B30-7	3280	2447	1.435m	1978-80
	212	U33C	3600	2686	1.435m	1969-75
	37	U30C	3280	2447	1.435m	1968-69
	10	U28C	3050	2275	1.435m	1966
	4	U28B	3050	2275	1.435m	1966
	3	U50D	5500	4103	1.435m	1964
	68	U25B	2750	2052	1.435m	1962-64
☐ Texas Utilities Service	2	B23-7	2490	1858	1.435m	1979-81
	1	U23B	2490	1858	1.435m	1975
	2	U18B	2000	1492	1.435m	1974
☐ Union Pacific System	60	C36-7	3940	2939	1.435m	1985
	58	B30-7A	3210	2394	1.435m	1980-82
	82	B23-7	2490	1858	1.435m	1978-81
	140	C30-7	3280	2447	1.435m	1977-80
	54	U23B	2490	1858	1.435m	1972-77
	40	U50C	5500	4103	1.435m	1969-71
	185	U30C	3280	2447	1.435m	1968-76
	21	U30B	3280	2447	1.435m	1967-72
	10	U28C	3050	2275	1.435m	1966
	23	U50D	5500	4103	1.435m	1963-65
	16	U25B	2750	2052	1.435m	1961-62
	4	UM20B	2000	1491	1.435m	1959

*Operated For N.J.D.O.T. and M.T.A.

WORLD USERS

INTERNATIONAL

Country / User	Quantity	Model	Gross HP	Gross KW	Gauge	Year
ALGERIA						
SNTF	25	U18C	1980	1476	1.435m	1976
ANGOLA						
C.F. de Luanda	5	U20C	2150	1604	1.067m	1969
	30	U20C	2150	1604	1.067m	1967
C.F. de Benguela	12	U20C	2150	1604	1.067m	1981-82
	12	U20C	2150	1604	1.067m	1974
	10	U20C	2150	1604	1.067m	1972
ARGENTINA						
Argentina State Ry.	4	U10B	1050	783	1.000m	1977
	45	U13C	1420	1059	1.676m	1962
	25	U13C	1420	1059	1.435m	1962
	11	U18C	2150	1604	1.676m	1960
	50	U12C	1420	1059	1.000m	1958-59
	20	U12C	1420	1059	1.435m	1958-59
	25	U18C	2150	1604	1.676m	1957
AUSTRALIA						
Mt. Newman Mining	6**	C36-7M	3950	2947	1.435m	1986
Queensland Railways	13**	U22C	2400	1790	1.067m	1982
Hamersley	3	C36-7	3940	2938	1.435m	1977
BANGLADESH						
Bangladesh Ry. Board	10	UM13C	1420	1059	1.000m	1964
BOLIVIA						
ENFE	9	U10B	875	522	1.000m	1977
	8	U20C	2150	1604	1.000m	1976-77
BOTSWANA						
Botswana Railway	12**	UM22C	2140	1604	1.067m	1981
	2	U10B	950	709	1.067m	1978
BRAZIL						
Acominas Gerais	3	UM10B	1050	783	1.600m	1979
Cia. Siderurgica Nacional	3	UM10B	1050	783	1.600m	1982
	11	UM10B	1050	783	1.600m	1975
	4	UM10B	1050	783	1.600m	1966
Campanhia Vale do Rio Doce	41	C30-7B	3350	2500	1.600m	1983-86
	6	U26C	2750	2052	1.000m	1981
COSIPA	2	U12C	1420	1059	1.600m	1961
	2	U6B	700	522	1.600m	1960
Ferrovia Paulista, S.A.	136	U20C	2150	1604	1.000m	1974-76
	10	U9B	1060	791	1.600m	1958-59
	5	U18C	2150	1604	1.000m	1958
	22	U12C	1420	1059	1.000m	1957
Rede Ferroviaria Federal, S.A.	10	U22C	2408	1796	1.000m	1985-86
	30	U20C	2150	1604	1.000m	1979
	30	U20C	2150	1604	1.600m	1979
	90	U23C	2500	1865	1.600m	1974-76
	105	U20C	2150	1604	1.600m	1974-75
	24	U20C	2150	1604	1.000m	1974
	80	U23C	2500	1865	1.600m	1972-73
	80	U10B	1050	783	1.000m	1971-72
	20	U6B	700	522	1.600m	1967
	54	U13B	1420	1059	1.000m	1963
	88	U5B	600	448	1.000m	1961-62
	20	U5B	600	448	1.600m	1961-62
	65	U8B	900	671	1.000m	1961
	30	U12B	1420	1059	1.000m	1958
	18	U12B	1420	1059	1.000m	1957-58
	3	U9B	1060	791	1.000m	1957
CHILE						
Anglo Lautaro Nitrate	3	U12C	1420	1059	1.067m	1957
Chilean State Railways	8	U13C	1420	1059	1.000m	1967
	4	U5B	600	448	1.676m	1962
	12	U5B	600	448	1.676m	1962
	4	U12C	1420	1059	1.000m	1960
	11	U9C	1060	791	1.000m	1957
Minera Sante Fe	4	U6B	700	522	1.000m	1960
	5	U9C	1060	791	1.000m	1957
CHINA						
Ministry of Railways	200	C36-7	4000	2984	1.435m	1985
	220	C36-7	4000	2984	1.435m	1983
COLOMBIA						
Cerrejon Coal Project	8	B36-7	3980	2968	1.435m	1982-84
Colombian Nat. Rys.	8	U18C	1650	1230	.914m	1984
	28**	U10B	1050	783	.914m	1973
	60**	U10B	1050	783	.914m	1969
	10	U20C	2150	1604	.914m	1964
	6	U13C	1420	1059	.914m	1961
	8	U8B	900	671	.914m	1961
	8	U6B	700	522	.914m	1959-60
	18	U12C	1420	1059	.914m	1958
COSTA RICA						
Cia. Bananera de Costa Rica	3	U6B	700	522	1.067m	1959
Ferrocarriles de Costa Rica	10	U11B	1100	821	1.067m	1978
	5	U10B	825	615	1.067m	1978
	6	U6B	700	522	1.067m	1971
	4	U6B	700	522	1.067m	1966
CUBA						
Occidentales Railways	10	U6B	700	522	1.435m	1959-60
	10	U12B	1420	1059	1.435m	1959
DOMINICAN REPUBLIC						
Central Rio Haina	1	U6B	700	522	1.435m	1959
GABON						
Comilog	3	U15C	1650	1231	1.067m	1975
	3	U15C	1650	1231	1.067m	1973
	6	U15C	1650	1231	1.067m	1971
	2	U15C	1650	1231	1.067m	1970
	1	U13C	1450	1082	1.067m	1968
	3	U13C	1450	1082	1.067m	1965
	10	U13C	1450	1082	1.067m	1960-64
OCTRA	6	UM22C	2335	1742	1.435m	1984
	6	UM22C	2335	1742	1.435m	1975
GREECE						
Hellenic Railway Organization	13	UM10B	1050	783	1.435m	1973
GUATEMALA						
Ferrocarriles de Guatemala	18**	U10B	1050	783	.914m	1971
	2	U8B	900	671	.914m	1962

**Built by General Electric Licensee

Tractive Effort: Improved alternator or DC generator, traction motors and gearing furnish high continuous tractive effort. SENTRY or improved CMR adhesion control systems allow full utilization of the engine's horsepower and greater usable tractive effort.

WORLD USERS

INTERNATIONAL

	QUANTITY	MODEL	GROSS HP	GROSS KW	GAUGE	YEAR
HONDURAS						
Ferrocarril Nacional de Honduras	2**	U10B	1050	783	1.067m	1972
	5	U10B	1050	783	1.067m	1967-68
	1	U10B	1050	783	1.067m	1966
Standard Fruit Co.	3	U6B	700	522	.914m	1971
	11	U5B	600	448	.914m	1963
United Fruit Co.	1	U10B	1050	783	1.067m	1970
	12	U6B	700	522	1.067m	1967
INDONESIA						
PJKA	18	U18C	1950	1454	1.067m	1983
	34	U18C	1950	1454	1.067m	1982
	30	U18A1A	1950	1454	1.067m	1982
	10	U18C	1950	1454	1.067m	1976
	11	U18A1A	1950	1454	1.067m	1976
	24	U18C	1950	1454	1.067m	1975
Pusri	4	U18C	1950	1454	1.067m	1975
IVORY COAST						
Sode Sucre	1	U6B	700	522	1.000m	1979
JAPAN						
Nippon Sharyo	1**	U10B	1050	783	1.067m	1970
JORDAN						
Aqaba Railway Corp.	18	U20C	2150	1604	1.050m	1980
	3	U18C	1950	1361	1.050m	1976
	10	U17C	1850	1380	1.050m	1973
Hedjaz Jordan Railway	5	UM10A1A	1050	783	1.050m	1974
KENYA						
Kenya Railways	10	U26C	2610	1946	1.000m	1987
	26	U26C	2610	1946	1.000m	1977
MALAYSIA						
Rompin Mining Co.	3	UM10B	1050	783	1.000m	1963
MEXICO						
Ferrocarril del Pacifico	4	C30-7	3280	2447	1.435m	1985
	22	C30-7	3280	2447	1.435m	1981
	15	C36-7	3940	2939	1.435m	1978
	10	U23B	2060	1536	1.435m	1975
	10	U36C	3940	2939	1.435m	1974
	4	U30C	3280	2447	1.435m	1971
	4	U30C	3280	2447	1.435m	1969
Ferrocarrilas Nationales de Mexico	30	C30-7	3280	2447	1.435m	1984
	30	C30-7	3280	2447	1.435m	1983
	114	C30-7	3280	2447	1.435m	1980-82
	112	B23-7	2060	1536	1.435m	1980-82
	15	C36-7	3820	2850	1.435m	1980
	30	C30-7	3280	2447	1.435m	1979
	25	C30-7	3280	2447	1.435m	1979
	21	C30-7	3280	2447	1.435m	1979
	10	B23-7	2529	1887	1.435m	1979
	25	C36-7	3940	2939	1.435m	1978
	29	C30-7	3280	2447	1.435m	1978
	1	C30-7	3280	2447	1.435m	1978
	17	U36C	3940	2939	1.435m	1974
	30	U23B	2490	1858	1.435m	1974
	87	U36B	3940	2939	1.435m	1973
	45	U18B	2000	1492	1.435m	1973
	10	UD18B	2150	1604	1.435m	1956
Sureste	3	B23-7	2529	1887	1.435m	1979

INTERNATIONAL

	QUANTITY	MODEL	GROSS HP	GROSS KW	GAUGE	YEAR
MOZAMBIQUE						
C.F. de Mozambique	10	U20C	2150	1604	1.067m	1984
	20	U20C	2150	1604	1.067m	1979-80
	10	U20C	2150	1604	1.067m	1979
	25	U20C	2150	1604	1.067m	1978-79
	22	U20C	2150	1604	1.067m	1973
	16	U20C	2150	1604	1.067m	1968-69
	16	U20C	2150	1604	1.067m	1966
Transzambesi Railway	5	U20C	2150	1604	1.067m	1972
NEW ZEALAND						
New Zealand Railways	6	U10B	1080	783	1.067m	1977
	34	U26C	2750	2052	1.067m	1974
	15	U26C	2750	2052	1.067m	1971
NICARAGUA						
Ferrocarril del Pacifico	3	U10B	1050	783	1.067m	1974
	6	U10B	1050	783	1.067m	1964
NIGERIA						
Nigerian Ry. Corp.	25	U18C	1950	1454	1.067m	1976
	6	U22C	2335	1741	1.067m	1975
	20	U18C	1950	1454	1.067m	1975
PAKISTAN						
Pakistan Railways	42	U20C	2150	1604	1.676m	1970
	23	U15C	1650	1231	1.676m	1970
	5	U6B	700	522	1.676m	1965
PANAMA						
United Fruit Company	1	U6B	700	522	.916m	1970
	2	U6B	700	522	.916m	1967
PERU						
Southern Peru Copper Corporation	16	U23B	2490	1858	1.435m	1974
PHILIPPINES						
Philippine National Railways	10	U15C	1650	1231	1.067m	1979
	10	U10B	1050	783	1.067m	1979
	20	U10B	1050	783	1.067m	1975
	5	U15C	1650	1231	1.067m	1973
	13	U10B	1050	783	1.067m	1965-66
	10	U6B	700	522	1.067m	1960
	30	U12C	1290	962	1.067m	1956
Philippine Ry. Co.	1	U6B	700	522	1.067m	1960

**Built by General Electric Licensee

Fuel Economy: The turbocharger, new fuel system, dynamic braking package, variable speed radiator fan, lower horsepower equipment blower, and standard low Idle offer greater fuel economy.

WORLD USERS

INTERNATIONAL

	QUANTITY	MODEL	GROSS HP	GROSS KW	GAUGE	YEAR
SOUTH AFRICA						
Anglo-Alpha Cement Company	1	U10B	1050	783	1.067m	1967
President Brand Gold Mine	2**	UM10B	1050	783	1.067m	1985
	1**	UM10B	1050	783	1.067m	1980
Buffelsfontaine	1	U10B	1050	783	1.067m	1974
Douglas Colliery	2	U26C	2750	2052	1.067m	1975
Impala Platinum	2	U10B	1050	783	1.067m	1974
ISCOR	2	UM10B	1050	783	1.067m	1979
	2	U26C	2750	2052	1.067m	1975
	8	U10B	1050	783	1.067m	1975
	6	U10B	1050	783	1.067m	1975
	20	U26C	2750	2052	1.067m	1974
	8	U10B	1050	783	1.067m	1974
	22	U26C	2750	2052	1.067m	1973
	3	U10B	1050	783	1.067m	1973
	13	U10B	1050	783	1.067m	1972
	4	U10B	1050	783	1.067m	1969
	3	U10B	1050	783	1.067m	1965
	12	U10B	1050	783	1.067m	1964
South African Rys.	24	SG10	1100	821	1.067m	1979
	30	U26C	2750	2052	1.067m	1978
	50	U15C	1650	1231	1.067m	1977
	50	U15C	1650	1231	1.067m	1974
	100	SG10	1100	821	1.067m	1973-74
	100	U26C	2750	2052	1.067m	1973
	20	UM6B	700	522	.610m	1973
	20	U15C	1650	1231	1.067m	1972-73
	50	U15C	1650	1231	1.067m	1972
	125	U26C	2750	2052	1.067m	1971-72
	25	U20C	2150	1604	1.067m	1969-70
	90	U20C	2150	1604	1.067m	1968-69
	10	U20C1	2150	1604	1.067m	1966
	65	U20C	2150	1604	1.067m	1965
	115	U18C1	2150	1604	1.067m	1959-61
	45	U12B	1420	1059	1.067m	1958
Union Corporation	1	SG15	1650	1231	1.067m	1977
	3	U10B	1050	783	1.067m	1974
	2	U10B	1050	783	1.067m	1973
SPAIN						
Ensidesa	3**	U10B	1050	783	1.668m	1974
	8**	U10B	1050	783	1.668m	1970-71
	6**	U10B	1050	783	1.668m	1968
Ferrocarriles de la Robla	10	U10B	1050	783	1.000m	1964-65
Puerto de Gijon	3**	U11B	1200	895	1.668m	1974
Rente	16**	UM10B	1050	783	1.668m	1969
	12**	UM10B	1050	783	1.668m	1968
	8**	UM10B	1050	783	1.668m	1966
	5**	UM10B	1050	783	1.668m	1965
Tajuna	5**	U11B	1200	895	1.668m	1974
SUDAN						
Sudan Railway Corp.	10	U20C	2150	1604	1.067m	1985
	10	U15A1A	1650	1231	1.067m	1981
	20	UM22C	2335	1742	1.067m	1974
	10	U15A1A	1650	1231	1.067m	1974

INTERNATIONAL

	QUANTITY	MODEL	GROSS HP	GROSS KW	GAUGE	YEAR
SYRIA						
Chemins de Fer Syriens	30	U17C	1850	1380	1.435m	1974
TANZANIA						
Tazara	5**	U30C	3210	2394	1.067m	1982
	9**	U30C	3210	2394	1.067m	1981
THAILAND						
Thailand State Rys.	10	UM12C	1320	985	1.000m	1966
	40	UM12C	1320	985	1.000m	1963
TUNISIA						
Compagnie des Phosphates de Gafsa	3	U10B	700	522	1.000m	1981
	14	U6B	657	490	1.000m	1975
	5	U6B	700	522	1.000m	1974
	4	U6B	700	522	1.000m	1965
	4	U5B	600	448	1.000m	1962
SNCFT	6	U22C	2500	1865	1.435m	1981
	14	U22C	2500	1865	1.000m	1981
	10	U10B	700	522	1.435m	1981
	18	U10B	700	522	1.000m	1981
	6	U6B	700	522	1.000m	1976
	3	U6B	700	522	1.435m	1976
TURKEY						
TCDD	40	U20C	2150	1604	1.435m	1964-65
	5	U18C	2150	1604	1.435m	1957-58
VIETNAM						
Vietnam Railway Sys.	25	U8B	900	671	1.000m	1964-65
	23	U8B	900	671	1.000m	1963
ZAIRE						
Gecamines	12	U10B	1050	783	1.067m	1974
ONATRA	5**	U15C	1650	1231	1.067m	1984
	8**	U15C	1650	1231	1.067m	1980
	4	U10B	1050	783	1.067m	1974
SNCZ	12	U15C	1650	1231	1.067m	1973
	15	U15C	1650	1231	1.067m	1971
	23	U15C	1650	1231	1.067m	1970
ZAMBIA						
Roan Consolidated Mines	1	U10B	1050	783	1.067m	1975
	2	U10B	1050	783	1.067m	1972
Roan Selection Trust	1	U10B	1050	783	1.067m	1975
Zambia Railways	10**	U20C	2150	1604	1.067m	1979
	10	U20C	2150	1603	1.067m	1975
	8	U20C	2150	1603	1.067m	1975
	6	U20C	2150	1604	1.067m	1973
	12	U15C	1650	1231	1.067m	1973
	23	U20C	2150	1604	1.067m	1970
	26	U20C	2150	1604	1.067m	1967
ZIMBABWE						
National Railway of Zimbabwe	10	U20C	2150	1604	1.067m	1966

**Built by General Electric Licensee

Training: GE Learning And Communication Center is dedicated to building locomotive productivity through technological training. Maintenance and repair procedures, as well as advanced methods of troubleshooting are taught through a combination of high-technology instruction and hands-on experience in specially equipped labs.

E60CP LOCOMOTIVE specification
section 1 ratings, weights, dimensions & supplies

MODEL — E60CP, 6000 HP, Six Motor Electric Locomotive

WHEEL ARRANGEMENT — C-C

RATINGS
- Primary voltages (±10%) .... 11,000 volts, 25 Hz
- 12,500 volts, 60 Hz
- 25,000 volts, 60 Hz
- Starting tractive effort .... 75,000 lbs.
- Continuous tractive effort .... 34,000 lbs.
- (68/38 gear ratio, 40" wheel diameter)
- Continuous rail horsepower 5,100
- Short time rail horsepower 9,800
- Maximum locomotive speed .... 120 MPH
- Speed-tractive effort curve 41H119081
- Speed-braking effort curve 41H115367
- Auxiliary power supply for train heat, and air conditioning .... 750 KW

DIMENSIONS
- Length inside coupler knuckles .... 71 ft. 3 in.
- Distance between bolster centers .... 45 ft. 0 in.
- Rigid wheel base .... 13 ft. 7 in.
- Width over handrails .... 10 ft. 5 in.
- Height over pantograph locked down .... 14 ft. 7.5 in.
- Minimum radius of curvature, locomotive alone .... 273 ft. (21°)
- Locomotive outline 41D717190
- Locomotive location of apparatus 41D717879
- Locomotive clearance diagram 41D717372
- Maximum locomotive weight .... 366,000 lbs.
- CG of car body (less trucks) .... 88.8 in. off the rail
- (including trucks) .... 66.4 in. off the rail
- Maximum displacement of CG of car body
  - 6 in. superelevation 3.41
  - maximum lateral 4.84

SUPPLIES
- Sand .... 56 cu. ft.

section 2 general

DESIGN The General Electric Model E60CP electric locomotive is especially designed and built to meet the requirements of modern passenger traffic. It receives single phase alternating current from an overhead contact wire, which is rectified to supply six railway type d-c traction motors, one geared to each driving axle.

OPERATION The locomotive is operated by a standard AAR master controller and by independent and automatic air brake valves conveniently located to permit operation with either end leading.

The direction of motion is controlled by a reverse lever. The throttle and reverse levers are interlocked to prevent operation of the reverser unless the throttle handle is in the OFF position.

SAFETY APPLIANCES All safety appliances are in accord with the General Electric Company's interpretation of current FRA regulations.

TESTING All component parts of the locomotive are given standard commercial tests before assembly on the locomotive.

Each complete locomotive is tested in accordance with Test Instruction 41A241577.

1. Control wiring is checked by observing the sequence of contactor and relay operation and by testing for continuity of circuit between terminals.

2. High potential tests of traction and control circuits are made in accordance with ANSI-IEEE standards for land transportation vehicles, ANSI C35.1/IEEE No. 11 for power circuits and ANSI C98.1/IEEE No. 16 for control and auxiliary circuits.

3. Air Brake tests are conducted to comply with FRA regulations and with G.E. instructions.

4. Each locomotive is track tested at the builder's plant to assure satisfactory operation.

section 3 platform and cab construction

UNDERFRAME	The welded underframe is made of rolled steel sections and plate.
	Space between these members is enclosed to form an air duct which distributes clean air throughout the locomotive.
COLLISION POSTS	Collision posts consisting of two main members are provided in both ends of the locomotive in the nose cabs. Ultimate shear value is not less than 600,000 lbs. in a longitudinal direction and not less than 300,000 lbs. in a lateral direction.
WEARPLATES	Renewable, wear-resistant hardened steel plates are applied to the centerplate, side bearing pads and draft gear housing.
COUPLERS	AAR type H bottom operated interlocking couplers with NC-390 rubber-cushioned draft gear and alignment control are provided at each end of the locomotive.
PILOT AND SIDE STEPS	A pilot is provided at each end of the locomotive. Side steps provide access to the operator's cabs. Foothold in pilot also provided to comply with FRA requirements.
LIFTING AND JACKING	Four jacking pads in combination with lugs for cable slings are provided in the side bolsters. Lifting lugs are also provided at both ends of the underframe for lifting one end at a time.
SUPER-STRUCTURE	The welded steel superstructure consists of operator's cabs and an equipment cab. The equipment load, tractive effort and buff loads are shared by the superstructure underframe. The superstructure carries part of the structural loads and is made up of stress members (changes to the superstructure should be avoided unless reviewed and approved by the locomotive builder).
OPERATOR'S CAB	The sides and roof of each operator's cab is insulated and steel lined. The floor, raised above the underframe, is covered with 1/2" thick Benelex.
	The cab has Lexguard windows in the front and sliding Lexan windows equipped with latches on each side of the cab. All other windows are fixed and are glazed with Lexan.
	Doors in both sides of the operator's cab provide access to the cab. The doors have windows, weather stripping and locks. Doors in the rear of the operator's cab provide access to the equipment cab.
EQUIPMENT CAB	The full width equipment cab encloses the power rectifiers, the a-c and auxiliary control, the transformer, the smoothing reactor, the equipment blower, the auxiliary power plant, and the air compressor. Access to this equipment is provided on both sides of the locomotive from the aisles. Removable hatches permit removal of the equipment where necessary.
VENTILATION	Air is drawn into the locomotive through "V" screens located in the roof. The air is filtered through self-cleaning air cleaners located under the blower. Clean air is delivered under pressure for equipment cooling, pressurization, and cab ventilation. Equipment cabs are pressurized with double filtered air.
BATTERY BOX	One battery box is provided on the right side of the locomotive.
DOOR LOCKS	James L. Howard Company, Model 2527 door locks are provided.

section 4 underframe and running gear

RUNNING GEAR The running gear of the locomotive consists of two, three-axle, lateral motion single centerplate swivel trucks.

Centerplate load is distributed by the cast-steel "floating bolster" to four rubber mounts which rest on the truck frame and provide controlled lateral motion. The truck frame is supported by alloy steel coil springs over the journal boxes.

Hydraulic shock absorbers damp vertical oscillation.

The truck frame is of cast steel.

AXLES Axles are solid forged steel, conforming to AAR material specifications.

WHEELS Class BR, multiple-wear, rolled steel wheels are 40-inch diameter with AAR tread and flange contour.

JOURNALS Journals are equipped with sealed grease-lubricated Timken GG roller bearings. Pedestal openings of the truck frame have renewable non-metallic wear plates.

CENTER PLATES Large center plates are equipped with liners and protected by dust guards.

SAFETY HOOKS Truck safety hooks are provided to minimize slewing in case of derailment and to permit the trucks to be lifted with the superstructure.

BRAKE EQUIPMENT Brake cylinders are mounted on the truck frames and operate fully equalized brake rigging, which applies one composition brake shoe to each wheel. Brake rigging is furnished with hardened steel bushings and adjustment is provided to compensate for wheel and shoe wear.

HAND BRAKE A hand brake is provided that applies the brakes to two wheels of the rear truck. The front truck is also equipped with the necessary levers to be interchangeable with the rear truck.

Sufficient braking power is developed to hold the fully loaded locomotive on a 3% grade when an actual force equivalent to 125 pounds is applied at the rim of the 20-inch diameter brake wheel.

section 5 propulsion equipment

TRANSMISSION SUPPLY
11,000 volt, 25 Hz, 12,500 or 25,000 volt, 60 Hz a-c power can be supplied to the transformer primary. Changes from one power supply to another are initiated by the operator. This is accomplished with the use of a motorized tap changer mechanism and associated control equipment. It permits remote operations allowing the power supply change to be made with the locomotive in motion. The transformer secondary supplies power to the bridge-connected silicon-type rectifiers. Rectified d-c power is supplied from the rectifiers to the traction motors through a smoothing reactor.

PANTOGRAPH
Two, spring-raised, air-lowered pantographs. Maximum extension 11 ft. per P.C. Std. 17MCP1A5.

A hand pump is provided for emergency use to supply air to lower the pantograph.

VACUUM BREAKER
A General Electric transportation duty robust construction vacuum circuit breaker type 17JP4B1 of 30 KV insulation class, it has two distinct operating rates with up to 8000 amp interrupting capability. Spring-charged mechanism provides rapid fault clearance and a pneumatic operator is used for normal circuit isolation.

LIGHTNING ARRESTOR
Two series connected 15 KV General Electric Alugard II, station class lightning arrestors provide protection against lightning and switching surges to the equipment aboard the locomotive. When the locomotive is operating from an 11 KV or 12.5 KV power source, one of these arrestors is shorted out.

TRANSFORMER
The main traction duty power transformer is a General Electric Design Class FOA and is of a welded case, three limbed core configuration, multiple secondary windings, forced Pyranol circulated and air blown cooling type. Construction, in keeping with the duty, includes firm winding bracings and internal core support arrangement as well as an external mounting configuration adequate to withstand the anticipated buffing forces and internal stresses resulting from abnormal electrical faults.

The 7960 KVA transformer capacity fully meets the traction requirements, the power for the various auxiliaries included on the locomotive and train auxiliary load. The transformer has two sets of secondary windings for propulsion power. Each set consists of three secondary windings which provide power through rectifying bridges to three parallel connected traction motors through a smoothing reactor. A seventh secondary winding (through a silicon rectifying bridge and reactor), provides the desired supply to auxiliary motors in the system.

RECTIFIER
The rectifier groups incorporate General Electric Thyristor Cells. Matched power cells in a presspak arrangement are cooled by filtered air. Two thyristor rectifiers, comprising one asymmetric bridge with two SCR's and two diodes, plus two symmetrical diode rectifier bridges, supply d-c power (through a smoothing reactor) to three traction motors each. Power for the equipment blower/air compressor motor is provided through an additional asymmetric thyristor/diode bridge.

MAIN SMOOTHING REACTOR
The main smoothing reactor (17EX57A1) contains a separate winding for each group of motors. The split C core construction has a resiliently mounted inner core. The coils are also resiliently mounted and are cooled by filtered air.

section 5 propulsion equipment (cont'd.)

TRACTION MOTORS Six General Electric GE 780B1 traction motors are furnished. The traction motors are direct current, series wound, and are separately ventilated by the cleaned-air system. The armatures are mounted in anti-friction bearings.

Motors drive through single-reduction spur gearing. They are supported by anti-friction bearings on the axles to which they are geared and by resilient nose suspensions on truck transoms.

CONTROL The locomotive is equipped with General Electric railway type double-end multiple unit control. The reverser, braking set up switch and tap switch contactors are electropneumatically operated. Variable voltage is provided by silicon rectifiers and diodes in combination with tap switches and firing circuits for the silicon rectifiers.

DYNAMIC BRAKING Equipment for braking the locomotive electrically, using the traction motors as generators and dissipating the electric power in resistors, using a braking resistor blower (5GY19A4), is provided.

BLENDED BRAKING An automatic combination of pneumatic and dynamic braking is provided. This function is performed through the use of a New York Air Brake blending valve, model No. N9461.

EP BRAKING A WABCO model CS-1 braking unit is provided to convert pneumatic brake signals to train line electric brake signals. This function serves to reduce brake application release times on the cars that make up the train (provided these cars are equipped with mating EP brake equipment).

PROTECTION Transformer primary instantaneous overcurrent relays, instantaneous overcurrent relays for each secondary winding, as well as a ground relay, are provided. Transformer over temperature protection and over pressure protection are provided. Thermal protection for the primary winding is provided electronically. A primary side line vacuum interrupter (17JP4B1) is provided for fault clearance. After tripping of this protective device, the pantograph lowers automatically.

Overspeed protection on the auxiliary motor is included. A primary lightning arrestor is provided.

Ground relay protection is provided for each of the two traction motor circuits. Ground relay protection is also provided for the auxiliary power system.

A motor operated circuit breaker (GE 41A241296 P1) is provided for the MA head end power plant. This breaker can be tripped from the operator's cab should the need arise.

Wheel slip correction equipment gives indication of slip or slide and provides automatic correction of wheel slips by a controlled momentary power reduction to the slipping axle.

74-VOLT D-C POWER SUPPLY One 74-volt, 15 KW static rectifier group supplies regulated d-c power for control and lights.

section 5 propulsion equipment (cont'd.)

STORAGE BATTERY A 32-cell lead-acid type, 280 ampere hour (Exide Model MS280) storage battery is furnished to supply power for control logic, lights and other auxiliaries. A battery charging smoothing reactor (17EX43A2) is used in the charging circuit.

TRACTION MOTOR CUTOUT The locomotive is equipped with two independent power circuits — one for each truck. Either truck can be cut out in an emergency. A traction motor cutout switch is located in the front operator's cab, mounted on the relay compartment door.

AUXILIARY MOTORS One Direct Current motor (5GY64A1) drives the compressor and blower. A smoothing reactor (17EX56A2) is used for the compressor/blower motor. A single phase a-c oil pump motor for transformer cooling is provided.

VENTILATING SYSTEM Filtered air is provided through self-cleaning air cleaners located in the equipment cab. Clean air is delivered under pressure for equipment cooling and pressurization. The rectifier cooling air is additionally cleaned by paper filters. A pressure switch senses lack of cooling air to the rectifiers and protects equipment from overheating due to loss of ventilation. A d-c motor driven auxiliary blower provides cooling air for the auxiliary power plant reactor and auto transformer located below the platform.

LOCKED WHEEL PROTECTION An axle alternator is installed on one end of each axle. A locked axle would be detected by the difference in wheel speed. The wheel slip buzzer and light would be activated. If the locomotive was in motoring or braking, there would be an automatic reduction of motoring or braking effort. Dead unit locked-axle protection is provided if the MU trainline jumper is properly installed.

OPERATOR'S STATION A standard AAR operator's control stand is furnished with motoring, dynamic braking and reverse levers, automatic and independent brake valves. Also a horn valve and bell ringer valve.

The following devices are located within easy reach of the operator at his station: headlights, gauge lights, sander and power source changeover switches, pantograph up, call and reset buttons and the window wiper valve. Warning lights are easily seen on the control stand.

The traction motor load meter and the air brake gages are mounted on the top of the control stand.

Control, light and set up switches are mounted on the rear cab wall.

BARCO SIS 400 speed indicating system is provided. The system consists of four speed indicators (mounted on both sides of each operator's cab), and a speed recorder mounted in the rear (No. 2) operator cab above the control compartment.

section 6 head end power

AUXILIARY POWER PLANT

A motor alternator (GDY59A1), smoothing reactor (17EX43A2) and control compartment (17KG339A1) will provide 750 KW electrical power for train heat and air conditioning. The electrical supply will nominally be 480 volt, 60 Hz three phase a-c, which will be trainlined with 2/3 pole receptacles.

The two bearing machine is self ventilated. A three-pole a-c motor operated circuit breaker is provided and can be tripped remotely. The motor driving the alternator will be powered by rectified a-c supplied from the auxiliary winding of the main transformer. Over the load range of the equipment, the frequency is regulated by phase controlling the a-c input to the d-c rectifier. Output of the alternator is regulated against load variations over the load range of the equipment, and this is accomplished by maintaining a constant frequency/voltage ratio.

In the event of line voltage outage, the MA unit will automatically restart upon restoration of the line voltage.

section 7 locomotive brakes

AIR BRAKES 26L brake schedule including self-lapping independent and special 26F control valve portions. The following features are also provided:

1. P2A valve and PC switch for service application.

2. Safety control from VAPOR Alertor, plus 1.

3. Independent brake valve pressure set for 55 psi.

4. Removable independent brake valve handle.

5. Removable automatic brake valve handle.

6. Three-position cutout cock on brake valve (out, frt., pass.) with "F" control valve set for graduated release.

7. A-1 charging valve and two No. 8 vent valves including automatic 30 second timed sanding for emergency applications.

8. A C1 suppression valve with a 17 psi permanent suppression feature utilizing an HB5D relay valve for temporary and permanent suppression.

9. WABCO CS1 electro-pneumatic interface units for electrically train-lined brake application and release signals.

10. Air brake and dynamic brake blending is automatic. Operation of the automatic brake initiates the blending on the locomotive. With brake pipe pressures that would ordinarily cause brake cylinder pressures on the range of 0 to 25 psi, there would be no brake cylinder pressure and the dynamic braking would be proportional to the amount of brake pipe pressure with full dynamic braking at 25 psi brake cylinder pressure. With brake pipe pressures that would ordinarily cause 26 to 55 psi brake cylinder pressure, there would be full dynamic braking and 1 to 30 psi brake cylinder pressure. Emergency position of the automatic brake handle will bypass the blending system and increase the brake cylinder pressure to 55 psi while maintaining full dynamic braking.

Above 80 MPH the brake cylinder pressure is reduced to 60% of the above values.

FOUNDATION BRAKES Four 12" x 4" truck frame mounted brake cylinders, V174 Cobra brake shoes. Brake cylinders at one end operate two brake shoes each, with a 3.14:1 lever ratio. The remaining two cylinders operate one brake shoe each, with a 1.57:1 lever ratio.

BRAKING RATIO Braking ratio at a 50 psi brake cylinder pressure is 28.55% for a locomotive weighing 366,000 pounds.

BRAKE PIPING Wrought steel pipe with AAR fittings are used. Generally, all piping 1/2" O.D. and under uses nominal size steel tubing with SAE fittings below deck and copper with SAE fittings above deck.

MAIN RESERVOIR Two 22-1/2" diameter x 76" steel reservoirs mounted beneath the underframe. Total capacity: 56,000 cu. in. Both No. 1 and No. 2 main reservoirs equipped with Salem 880 filters and the main reservoir drains using a Salem 872 electronic timer.

section 7 locomotive brakes (cont'd.)

AIR COMPRESSOR One, d-c motor driven, single-stage, positive displacement, oil flood lubricated (New York Air Brake Company Model N-9479) rotary compressor delivering 252 cu. ft. per minute at 2550 RPM input. The system includes a 2 stage air/oil separator (Kargard Model WP150), and an air to oil heat exchanger. Compressor air intake is doubly filtered by the inertial filter and the paper filters.

Two operating speeds are provided. The lower speed is 1275 RPM providing 125 cfm of air flow at 140 psi. The higher speed is 2550 RPM providing 252 cfm of air flow at 140 psi.

Electric air compressor governor adjusted to maintain reservoir pressure between 130 and 140 psi.

SAND CAPACITY Four sand boxes with a total capacity of 56 cu. ft.

Sanding systems are controlled electrically by a Salem 500 BS control switch. Manual sanding switch or automatic sanding in power operates eight single line Salem 277-2 sand traps, four traps for forward movement and four traps for reverse movement. A separate switch is provided for lead axle sanding only. Sand trap cutoff valves are provided. Outside access is provided for trap maintenance.

EMERGENCY VALVE Conductor's emergency (WABCO) valve is provided on the back of the control stand.

GAUGES AND TEST FITTINGS Salem 4-1/2" duplex air gauges with gauge test fittings are standard. Test fitting is also supplied at compressor unloader switch.

SIGNAL WHISTLE An electrically controlled conductor's air signal whistle is provided in each operator's cab.

section 8 locomotive accessories

SEATS	Three, slide rail mounted seats equipped with back rests and arm rests. Cushioned arm rests are provided at side windows. Engineer's seat: Car and Coach Co. No. 1555109. Fireman's seat: Car and Coach Co. No. 1114010.
SUN VISORS	Four, adjustable-type.
SUPPLY BOXES	In each operator's cab.
TOILET	A VAPOR Newmatic toilet model 17360534 is provided. Toilet tissue holder West Model 1119 also provided.
TRAIN COM-MUNICATION	Motorola Micor radio handset is installed in each operator's cab. Also a system for communication with the train crew and passengers is provided (Safetron Model 074-5670-01).

AMTRAK provided the following radio equipment for G.E. installation:

1 Model TLN1007C Mounting Rack
2 Model TCN1099A Control Head
1 Model TED611DA Antenna - 4" low profile

WATER COOLER	A floor mounted Vortacool Model No. VR770FWS water cooler in each operator's cab.
CUP DISPENSER	An Ajax Model E185 cup dispenser is provided in each cab.
WINDOW WIPERS	Two Sprague Jumbo Model M100, air-operated mounted on the front windows of each operator's cab.
WINDSHIELD WINGS	Wind deflectors, one at the front of side windows on both sides of each operator's cab. They include a bottom rear view mirror.
WINDSHIELD WASHERS	One Sprague Sprakleer No. 4 windshield washer with six nozzles is provided on each end to wash the windshields.
NUMBER BOX	Number boxes are provided on each end.
CLASSIFICA-TION LIGHTS	Classification lights are provided on each end.
MU TRAINLINE & RECEPTACLES	Following trainline functions are provided for:

MU
Communication
Three phase power

WASTE CONTAINER	A waste container, West Model No. 10, is provided in each cab.

section 8 locomotive accessories (cont'd.)

COMPOSITE CAB SIGNAL, TRAIN SPEED CONTROL AND AUTOMATIC TRAIN STOP

Complete AMTRAK special design US&S Schedule 358 composite continuous cab signal and train speed control including the following equipment in each operator's cab:

1. A "mushroom" type electric acknowledging switch located in control stand below air operating valve for bell.

2. A red indicating light located in a separate box mounted on the forward edge of the control stand, labeled, "Overspeed" is provided to indicate the following:

    a. Flashing for overspeed condition not suppressed or downward change in cab signal indications which have not been acknowledged.

    b. Steady red for overspeed condition suppressed and acknowledged and will remain until speed requirement is met.

    c. Steady red if penalty brake application occurs and will remain until train has stopped and entire system is reset and recovered.

3. A white indicating light located in a separate box, labeled "S.C., not Oper." provided to indicate the following:

    a. Mode selector switch cut out.

    b. Speed control cut out switch cut out.

    c. Speed control switch in off position.

    d. Air supply to EP valve cut out (accompanied by addition of pneumatic switch in air supply line).

4. A blue indicating light located in a separate box labeled "No Speed Signal", indicates no output from axle generator.

5. Speed control "On-Off" switch provided on equipment shelf.

6. Speed control "In-Out" switch located in each control console.

7. Cab signal indicator mounted on center windsheild post.

8. 32 volt 7.5 amp, d-c static converter provided for cab signal and train stop power supply.

9. Axle generator for speed control mounted on Left No. 5 journal box.

BELL Two, stationary, cast iron, with air-operated ringer and operating valve.

CAB HEATER One, forced air, electric type with ducts to provide front window defrosting in each cab.

CLOTHES HOOKS Three, folding clothes hooks in each operator's cab. Adams and Westlake Model 1040.

section 8 locomotive accessories (cont'd.)

FIRE EX-TINGUISHERS — Two, twenty pound carbon dioxide, one in each operator's cab, Norris Industries Model 20-CO. Two, thirty pound dry chemical, in the equipment cab, SAFE-T-METER, Model 4050.

FIRST AID KIT HOLDER — Mounted on rear bulkhead of No. 2 end of operator's cab.

FLAG AND MARKER BRACKETS — Four brackets, two at each end of the locomotive (Adams and Westlake No. P232390).

FUSE HOLDER — Spare fuses are stored in close proximity to point of application.

HEADLIGHTS — At each end of the locomotive, each consisting of two, 200 watt, 30-volt, sealed-beam lamps. Dimming control is provided. A dual beam Gyralite, Translite, Model No. 20585-ACF, will also be provided with a white light during normal operation and a red light in case of an emergency.

HORN — One, five tone horn (Nathan, Type P) mounted on each end of the locomotive.

LIGHTING — Lights and outlets located as follows:

- Ceiling cab lights
- Equipment cab lights
- Gage lights
- Step lights by each ladder
- Outlet receptacles in equipment cab and operator's cab
- Number lights at both ends
- Classification lights on both ends

PANTOGRAPH POLE — One, for raising or lowering the pantograph in emergency.

SAFETY CONTROL — VAPOR Alertor Plus 1 safety deadman control.

SANDERS — Eight, pneumatically operated, arranged to sand ahead of the lead wheels of each truck in each direction.

section 9 painting

INTERIOR — Gray enamel, Gliddens No. 447-E-0110

UNDERFRAME & RUNNING GEAR — Blue/Black (acrylic), Dupont No. 939-67391

EXTERIOR — Color and design as shown on G.E. Drawing 41B517155. Scotchlite, Cat. No. BPM-1 material provided by 3M Co. is used on the exterior.

section 10 performance characteristics

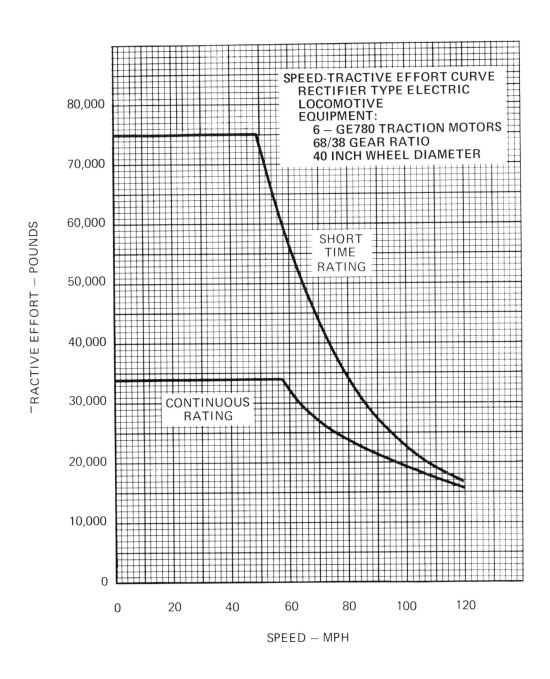

-193-

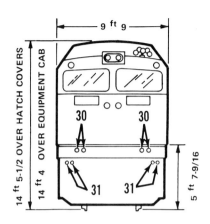
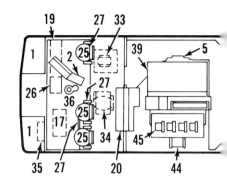
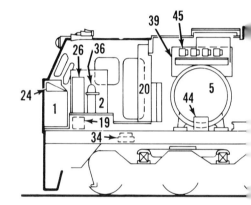
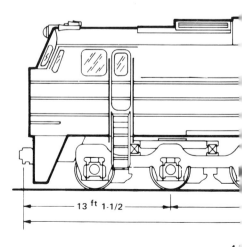

1. Sand Box
2. Control Stand
3. Control Compartment
4. Toilet
5. Motor Alternator Set
6. Dynamic Braking Resistor
7. Hand Pump
8. Equipment Blower
9. Blower & Comp. Motor
10. Air Compressor
11. Oil Sump
12. Smoothing Reactor
13. Rectifiers
14. Equipment Air Filters
15. Paper Filters
16. Transformer
17. Cab Heater
18. Aux. Motor Smoothing Reactor
19. Air Brake Equipment
20. Relay Compartment
21. Air Reservoir
22. Pantograph
23. Vacuum Circuit Breaker
24. Sand Filler
25. Cab Seat
26. Refrigerator
27. Cab Heater Register
28. Battery
29. Hand Brake
30. MU Receptacle
31. Power Receptacles
32. Comp. Control Panel
33. Battery Chg. Sm. Reactor
34. Battery Chg. Trans.
35. E.P. Air Brake Equipment
36. Fire Ext.
37. Wheel Slip Trans.
38. Telephone Interference Equipment
39. M.A. Control Compt.
40. Compressor Cooler
41. Auxiliary Blower
42. Auto-Transformer
43. M.A. Smothing Reactor
44. Control Trans.
45. M.A. Rectifiers

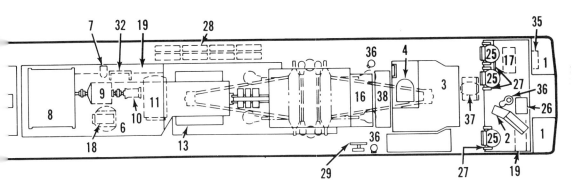
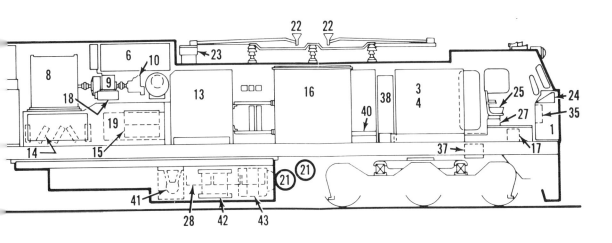
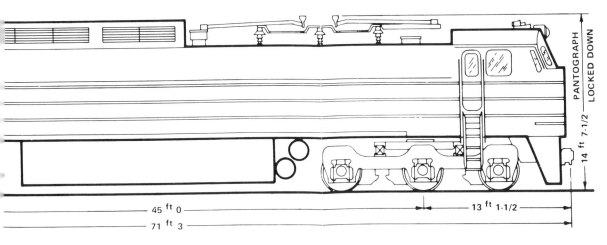

GEARCASE   4-5/8 UNDER MOTOR

-195-

section 10 performance characteristics (cont'd.)

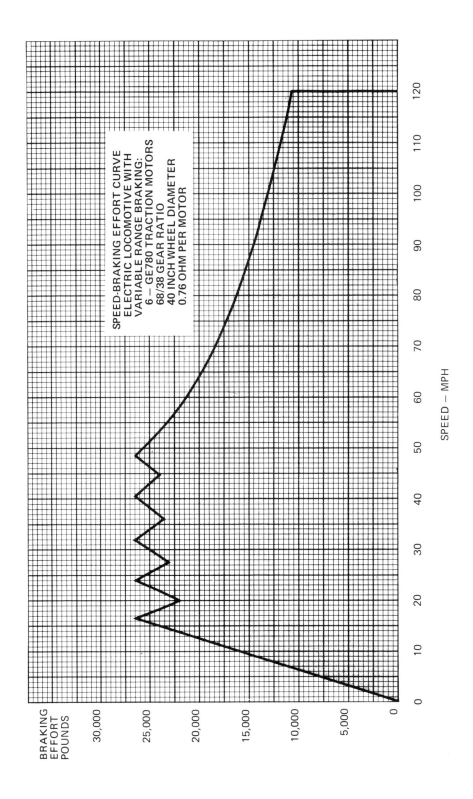

General Electric E60C General Characteristics

RATINGS		
Nominal voltage		25,000 v
Maximum voltage		27,500 v
Minimum voltage (continuous)		19,000 v
Minimum voltage (short time)		17,500 v
Continuous rail power		4,400 kw

	English	Metric
Continuous tractive effort (83/20 gearing, 40 in. 1016 mm wheels)	82,051 lbs	37,218 kg
Maximum starting tractive effort	116,895 lbs	53,023 kg
Maximum locomotive speed	69 mph	110 km/h

WEIGHTS		
Minimum locomotive	330,690 lbs	150,000 kg
Minimum per driving axle	55,115 lbs	25,000 kg
Maximum locomotive	431,881 lbs	195,900 kg
Maximum per driving axle	71,980 lbs	32,650 kg

Weight subject to manufacturing tolerance of ± 2%.
Modifications may increase weight.

WHEEL ARRANGEMENT	C-C	Co-Co
CAPACITY		
Sand	60 cu ft	1,700 liters

DIMENSIONS	English	Metric
Length inside knuckles	71 ft	21.609 mm
Height over pantograph locked down	16 ft 1 in	4,877 mm
Width over platform	10 ft	3,023 mm
Clearance under gear case	4⅓ in	111 mm
Minimum radius of curvature (locomotive alone)	262 ft	80 m
Track gauge	56½ in	1.435 m

MAJOR EQUIPMENT
Traction motors .... Six GE-752
Control .... Double end, multiple unit, electro-pneumatic type
Transformer .... 10C Mineral oil-filled
Rectifiers/Thyristors .... Silicon type
Traction motor blower .... D-C motor driven type
Compressor .... One, reciprocating compressor, D-C motor driven
Air brake schedule 26L
Pantographs .... Two, air-raised, spring-lowered

JOURNAL BEARINGS
Type .... Grease/tapered roller
Class and size .... GG (6⅞)

Location of Apparatus

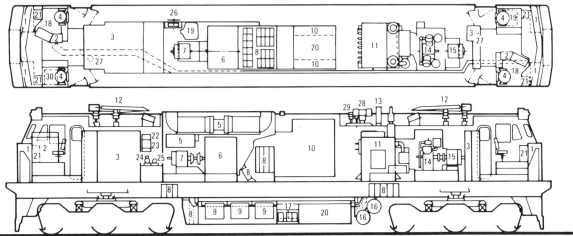

LEGEND

1. Sand Box
2. Control Stand
3. Control Equipment
4. Seat
5. Dynamic Braking Equipment
6. Equipment Blower
7. Blower Motor
8. Air Filters
9. Smoothing Reactor
10. Rectifiers
11. Transformer
12. Pantograph
13. Lightning Arrester
14. Air Compressor
15. Compressor Motor
16. Air Reservoir
17. Batteries
18. Air Brake Equipment
19. Auxiliary Smoothing Reactor
20. Harmonic Filter
21. Heater
22. Radio
23. Radio Filter
24. Auxiliary Air Compressor
25. Auxiliary Compressor Motor
26. Handbrake
27. Fire Extinguisher
28. Vacuum Circuit Breaker
29. Selector Switch
30. Auxiliary Transfer

General Electric E25B General Characteristics

	English	Metric
RATINGS		
Primary Voltage	25,000 kv, 60 Hz	25 kv
Locomotive continuous hp rating at rails	2,110 hp	574 kw
Maximum continuous tractive effort	55,000 lbs	24,950 kg
(74/18) gear ratio, wheel diameter	40 in	1,016 mm
Maximum tractive effort at starting	80,000 lbs	36,290 kg
Maximum locomotive speed	70 mph	112 km/hr
WEIGHTS		
Total locomotive (fully loaded)	280,000 lbs	127 T
Per driving axle (fully loaded)	70,000 lbs	31,750 kg
Weight subject to manufacturing tolerance of ±2%. Modifications may increase weight.		
WHEEL ARRANGEMENT	B-B	Bo-Bo
DIMENSIONS		
Length inside knuckles	64 ft 2 in	19.55 m
Height over pantograph locked down	16 ft	4.88 m
Minimum pantograph operating height	16 ft 8 in	5.08 m
Width over handrails	10 ft 3⅛ in	3.13 m
Minimum radius of curvature, locomotive alone	150 ft (39°)	45.70 m

MAJOR EQUIPMENT
- Traction motors .... Four GE752
- Control .... Single station, single unit, electro-pneumatic type
- Transformer .... Oil-Filled
- Rectifiers .... Silicon type
- Traction motor blower .... D-C motor driven type
- Compressor .... One rotary
- Air brake schedule 26L
- Pantograph .... Air-raised, spring-lowered

JOURNAL BEARINGS
- Type .... Grease/tapered roller
- Class and size .... GG (6⅞)

Location of Apparatus

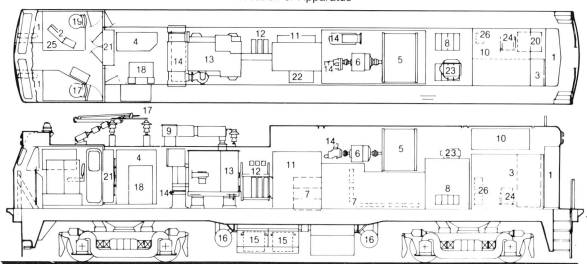

LEGEND

1. Sand Box
2. Control Stand
3. Control Equipment
4. Air Brake Equipment
5. Equipment Blower
6. Blower Motor
7. Air Filters
8. Batteries
9. Vacuum Circuit Breaker
10. Dynamic Braking Resistor
11. Rectifiers
12. Transformer Equipment
13. Transformer
14. Air Compressor Equipment
15. Smoothing Reactor
16. Air Reservoir
17. Pantograph
18. Air Conditioner
19. Seat
20. Wheel Slip Control
21. Relay Compt.
22. Automation Equip.
23. Aux. Motor Smoothing Reactor
24. Aux. Transformer
25. Fire Extinguisher
26. Battery Charg. Sm. Reactor

Reprinted from the 1984 Car and Locomotive Cyclopedia, Copyright, Simmons-Boardman Publishing Corp.

General Electric E42C General Characteristics

	English	Metric
RATINGS		
Primary voltage	25 kv, 60 Hz	25 kv
Locomotive continuous hp rating at rails	3,750 hp	2,800 kw
Starting tractive effort (91/20 Gearing)	59,980 lbs	27,207 kg
Continuous tractive effort (91/20 Gearing)	44,320 lbs	20,104 kg
Maximum locomotive speed (91/20 Gearing)	68 mph	109 km/hr
Wheel diameter	36 in	914 mm
WEIGHTS		
Nominal locomotive (fully loaded)	202,860 lbs	92,017 kg
Per driving axle (fully loaded)	33,810 lbs	15,336 kg
Weight subject to manufacturing tolerance of ± 2%. Modifications may increase weight.		
WHEEL ARRANGEMENT	C-C	Co-Co
CAPACITY		
Sand	12 cu ft	340 liters

	English	Metric
DIMENSIONS		
Length inside knuckles	55 ft 11¾ in	17.05 m
Height over pantograph locked down	13 ft 5⅓ in	4.10 m
Width over handrails	9 ft 9 in	2.97 m
Minimum radius of curvature, locomotive alone	186 ft	56.70 m
MAJOR EQUIPMENT		
Traction motors		Six GE-761
Control		Double end, multiple unit, electro-pneumatic type
Transformer		10C Mineral oil-filled
Rectifiers/Thyristors		Silicon type
Traction motor blower		D-C motor driven type
Compressor		One, rotary compressor, D-C motor driven
Air brake schedule		26LA
Pantographs		Two, air-raised, spring-lowered
JOURNAL BEARINGS		
Type		Grease/tapered roller
Class and size		E (6 x 11)

Location of Apparatus

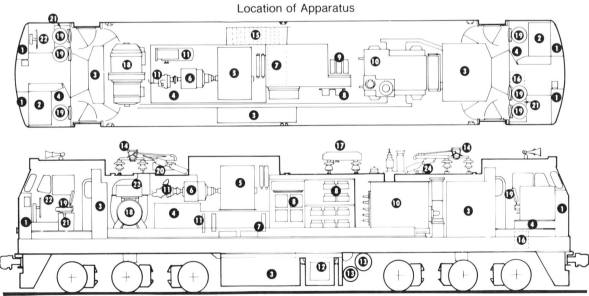

LEGEND

1. Sand Box
2. Control Stand
3. Control Equipment
4. Air Brake Equipment
5. Equipment Blower
6. Blower Motor
7. Air Filters
8. Rectifiers
9. Transformer Equipment
10. Transformer
11. Air Compressor Equipment
12. Smoothing Reactors
13. Air Reservoirs
14. Pantograph
15. Batteries
16. Auxiliary Transformer
17. Vacuum Breaker
18. M.A. Set (Modification)
19. Seat
20. Aux. Motor Smoothing Reactor
21. Locker
22. Handbrake
23. Aux. Air Compressor and Motor
24. Batt. Chg. Sm. Reactor

ELECTRIC MINE LOCOMOTIVE SPECIFICATIONS

Explore the results of one hundred years of mine locomotive design refinement

General Electric fully-engineered standard mine locomotives provide maximum value for capital invested. These standard models are easier to produce, can be shipped quicker, permit better parts support, better quality control, and greater resale value. They employ all of the major features required for maximum performance, convenience, safety, comfort, and compliance with regulatory requirements as interpreted and understood by the General Electric Company. In addition, they can be easily equipped with a wide variety of pre-engineered standard options.

General Electric standard, quantity-produced, modern design, Class H rated, type GHM traction motors assure high tonnage hauling efficiency and operating dependability.

This unsurpassed traction motor performance and precise ten-step cam contact controller combine to permit smooth, fast acceleration and high continuous duty. The maximum weight of the locomotive is utilized for operation close to the adhesion limit without wheel slipping.

Safety engineered suspension and braking systems permit maximum hauling speeds essential to modern mine operation.

Dynamic braking provides safe positive speed control on long downgrades and during deceleration. This standard feature reduces brake shoe maintenance and provides additional braking safety.

Rounded end frames, totally fire-resistant materials including cable insulation and seat covering, safety grounding of electrical equipment, "deadman" switch, and optional fire suppression system are among the many safety features introduced by General Electric.

Durable components, easy inspection access, and time-saving features such as lynch pin and dowel cover fastenings minimize maintenance and assure maximum utilization for haulage and/or supplies delivery.

Wire and cable are neatly arranged, securely clamped and protected against mechanical damage. Accelerating/braking resistors feature a special alloy ribbon element, edgewise wound, and supported on "floating" porcelain insulators.

Standard General Electric mine locomotives are available for either 250 or 500 nominal volt systems, in two or four-axle, high horsepower models for track gages of 42, 44 or 48 inches. Wheel/axle and motor sets are interchangeable between the 20-ton, two-axle, and 35-ton, four-axle models and between the 25-ton, two-axle, and 50-ton, four-axle models. Controllers are of the same basic design and most of the other control devices are identical on all units.

These four basic standard models will provide the optimum motive power for most mining needs. Numerous pre-engineered, standard options are available. Special customer designs for 36-inch gage and/or restrictive dimensions, as well as specialized features can be provided when required. Basic equipment and systems will remain as standard as practical.

General Electric's comprehensive renewal parts program includes detailed illustrated parts catalog, individually prepared spare parts recommendations, and an adequate stock of parts available from the factory warehouse and conveniently located regional parts centers.

1. General Electric GHM traction motor.
- BOX-TYPE FRAME CONSTRUCTION eliminates wear between frame and heads, insures long bearing life and minimizes maintenance.
- ONE-PIECE FORGED STEEL GEARS AND PINIONS feature long and short addendum teeth, accurate machining and heat treatment.
- ROLLER MOTOR AXLE SUPPORT BEARINGS minimize maintenance and power demand.
- EXTRA LARGE COPPER SECTIONS AND CLASS H INSULATION assure maximum hauling capacity with lowest power consumption, minimum motor inspection and maintenance, maximum brush life and flashover protection. All motors are double vacuum pressure impregnated for maximum moisture resistance and heat transfer.

2. Four-axle locomotives are available in 35 and 50-ton models.
3. Two-axle locomotives are available in 20 and 25-ton models.

Standard Features

FRAME — Rolled steel plate side frames, outside of wheels. Semi-circular end frames of steel plate formed and welded integral with side frames offer maximum operator protection.

COUPLERS — Willison automatic 10½ to 14½-inch heights. Shank length is 33 inches on two-axle models and 42 inches on four-axle models.

RUNNING GEAR

WHEELS — Steel-tyred with rolled steel centers. NEMA tread contour.
AXLES — Quenched and tempered carbon steel. AISI 1045.
JOURNALS — Pedestal type, renewable shims and frame guides. Two Timken AP type tapered roller bearings per box, grease lubricated.
SPRING RIGGING — Two-axle models - grouped helical springs over journals. Four-axle models - helical springs with compensating side equalizers.

BRAKES – AIR OPERATED

RIGGING — Fully equalized and adjustable for wear.
BRAKE SHOES — Flanged, two-piece cast iron shoe with hardened inserts.
HAND BRAKE — Hand wheel and square thread screw. Dash mounted, chain locked against unwinding.
POWER BRAKE — Straight air brakes with SA-26 self-lapping brake valve, ASME certified reservoir, duplex air gage, safety valve and railway type brake cylinders. Emergency air reservoir on trucks of four-axle units with separate pushbutton valve.

COMPRESSOR — Motor belt driven, continuous operation, controlled by unloader valve. Type 2 AVS, displacement 25 cfm, operating pressure 90-105 psi.

TRACTION MOTORS — Type GHM series wound, commutating pole, railway traction motors, Class H insulation, box-type frame with rubber wafer nose suspension and safety bar. (One motor per axle.)

BLOWERS — Single two-speed blower for four-axle models. Two individual blowers for two-axle models.
ARMATURE BEARINGS — Roller type.
AXLE BEARINGS — Anti-friction roller bearings, grease lubricated. Ground brush on each motor.
GEARING — Single reduction spur gears, heat treated, long and short addendum teeth. Gear ratio: 4:5:1
GEARING BOX — Fabricated steel plate, welded construction with renewable ¼-inch wear shoe.

CONTROL — Resistance/contactor control operating on battery voltage; battery charging from alternator conversion.

CONTROLLER — Model 17KC62 cam contact provides 10 motoring steps and 7 dynamic braking steps.
REVERSER — Separate, electro-pneumatic.

CONTACTORS — Railway type electro-magnetic or electro-pneumatic.
RESISTORS — EWF edgewise wound, alloy steel ribbon units on "floating insulators".
PROTECTIVE DEVICES — Overload relays in main trolley and each motor circuit, inoperative during dynamic braking reset when tripped in "off" position of controller. No voltage relay. Four-second time delay relay in auxiliary equipment circuit.
CONTROLLER AIR RESERVOIR — For emergency motoring and braking control.
BATTERY — Heavy-duty, lead-acid, 30 VDC, 5-3 cell trays.
CIRCUIT PROTECTION — Circuit breakers for control and lighting. Fuses for auxiliary motors.

TROLLEY

POLARITY — Specify negative or positive.
TYPE — Reversible, wood pole, swiveling and removable base.
POLE LENGTH — 5 feet.
LOCATION — (Specify) Right/left (in operator's cab on four-axle models).

STANDARD ACCESSORIES

HEADLIGHTS — Two per end, sealed beam, 150-watt, 30-volt operated from battery.
SANDERS — Air operated, all wheel sanding.
WARNING DEVICES — Foot gong and air horn with manual valve.
SAFETY CONTROL — Deadman foot switch, with adjustable time delay, removes power and actuates emergency air brakes. Customer to advise if emergency air brake feature not desired.
LOW PRESSURE ALARM — Audible signal, activated at 70 psi or below.
LOW PRESSURE LOCKOUT — Prevents starting until air pressure is normal.

STANDARD INSTRUMENTATION

BATTERY VOLTMETER — Monitors charging system (illuminated).
TRACTION MOTOR AMMETER — One motor circuit only (illuminated).
AIR GAGE — Duplex gage to read main reservoir and brake cylinder pressure (illuminated).

MISCELLANEOUS

EQUIPMENT SPACE — Space available for communications radio.
REFLECTORS — 3-inch, red, on all four corners.
SEAT — Two-person, bench-type, padded.
LIFTING ARRANGEMENTS — Provided.
PAINT — High visibility yellow with black lettering and numbering.
CAB ARRANGEMENT — Specify right or left hand.

Standard Optional Features

- Hooded couplers (top and bottom).
- Manual remote uncouplers.
- Air uncoupler No. 1 end and lever for No.2 end.
- Air uncouplers both ends (four-axle units only).
- Manual trolley reactor (air operated).
- Automatic trolley reactor (air operated).
- Blower indicating lights.
- Trolley voltmeter or extra motor ammeter (mutually exclusive).
- MU Control (25-ton units only).
- Air system alcohol injector.
- Windshields (one or both directions).
- Speedometer (magnetic pickup from traction motor gearing).
- Extra trolley socket (for spare pole).
- Resistance cap heater (with switch and safety ground).
- Draft gear (two-axle models only).
- Operator entrance cutout.
- Ansul fire suppression system.
- Reproducible drawings.
- Truck construction for rough track.

Two-Axle Standard Models

RATINGS		
Model	LM-2C20	LM-2C25
Specification number	RY-24881	RY-24882
Weight, tons (± 2% tolerance)	20	25
Rated voltage (VDC nominal)	250/500	250/500
Maximum drawbar pull (lbs)	12,000	15,000
Rated drawbar pull (lbs) (dbp)	10,000	13,500
Speed at rated dbp (mph)	10	10
Motor horsepower - continuous	280	376
Motor horsepower - one hour	320	432
Ventilation (cfm)	400	500
Axles	2	2
DIMENSIONS (Inches)		
Track gages available	42/44/48	42/44/48
Maximum width	80	84
Maximum height (over frame)	36	41
Maximum height (over trolley/base)	41	47½
Maximum length (over end frames)	276	304
Wheel base	108	108
Wheel diameter	31	31
Minimum clearance (under gear box)	2½	1⅞
Minimum curve radius (feet, locomotive alone)	45	45

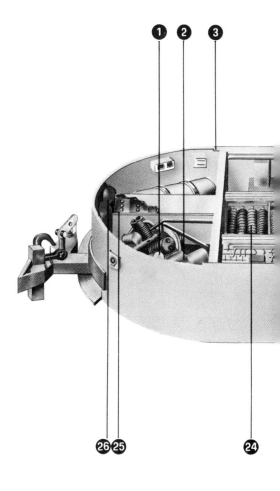

① Motor belt driven compressor

② Battery-charging alternator with integral rectifier and regulator

③ Integrally-welded steel plate frame construction

④ Fully-equalized, wear adjustable, air operated brakes

⑤ General Electric, high-efficiency, series wound, type GHM, commutating pole railway traction motor

⑥ Grease-lubricated, anti-friction, roller-type motor axle bearings

⑦ Pedestal-type, dual-tapered cartridge roller bearing journal boxes

⑧ Air operated sanders at all wheels

⑨ High-visibility illuminated instrument panel

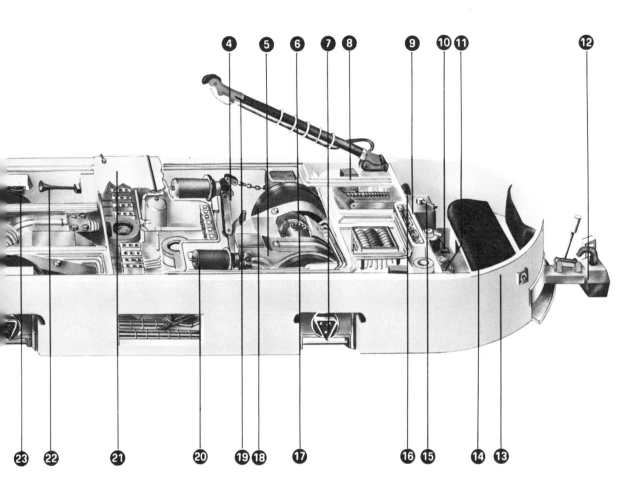

- ⑩ Deadman foot switch
- ⑪ Foot gong
- ⑫ Willison automatic couplers
- ⑬ Rounded end frames
- ⑭ Two-person, bench-type, padded seat and back rest
- ⑮ Cam contact controller with 10 motoring steps and 7 dynamic braking steps
- ⑯ Bulkhead mounted chain locked hand brake
- ⑰ Quenched and tempered axles
- ⑱ Steel-tyred wheels
- ⑲ Reversible, swiveling, removable, wood trolley pole with slider shoe
- ⑳ Multiple cylinder air brake system with standard railroad brake components
- ㉑ Easy access equipment inspection covers
- ㉒ Air horn
- ㉓ Springs over journal boxes
- ㉔ Accelerating/dynamic braking resistors
- ㉕ Heavy-duty lead-acid battery for control and lights
- ㉖ Four sealed beam, 150-watt, 32-volt headlights

Four-Axle Standard Models

RATINGS		
Model	LME-2C2C35	LME-2C2C50
Specification number	RY-24883	RY-24884
Weight, Tons (± 2% tolerance)	35	50
Rated voltage (VDC nominal)	250/500	250/500
Maximum drawbar pull (lbs)	21,000	30,000
Rated drawbar pull (lbs) (dbp)	18,500	25,000
Speed at rated dbp (mph)	10	10
Motor horsepower - continuous	560	752
Motor horsepower - one hour	640	864
Ventilation (cfm)	500	500
Axles	4	4
DIMENSIONS (Inches)		
Track gages available	42/44/48	42/44/48
Maximum width	82	88
Maximum height (over frame)	41	41
Maximum height (over trolley base)	47½	47½
Maximum length (over end frames)	433½	438
Wheel base	77	77
Wheel diameter	31	31
Minimum clearance (under gear box)	2½	1⅞
Minimum curve radius (feet, locomotive alone)	35	35
Truck centers	216	216

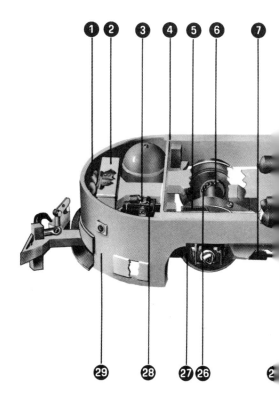

1. Four sealed beam, 150-watt, 32-volt headlights
2. Heavy-duty lead-acid battery for control and lights
3. Motor belt driven compressor
4. Integrally-welded steel plate frame construction
5. Accelerating/dynamic braking resistors
6. Quenched and tempered axles
7. Fully-equalized, wear adjustable, air operated brakes
8. Emergency brake reservoir
9. Two two-axle fabricated swivel trucks with lubricated center plates
10. General Electric high-efficiency, series wound, type GHM, commutating pole railway traction motor

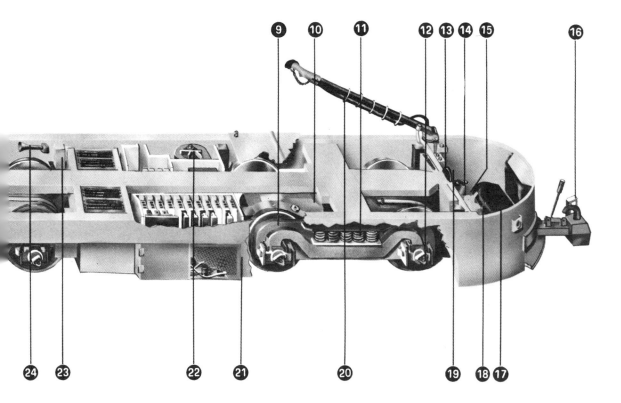

- ⑪ Equalized bar spring suspension
- ⑫ Pedestal-type, dual-tapered cartridge roller bearing journal boxes
- ⑬ High-visibility illuminated instrument panel
- ⑭ Cam contact controller with 10 motoring steps and 7 dynamic braking steps
- ⑮ Deadman foot switch
- ⑯ Willison automatic couplers with rubber wafer draft gear
- ⑰ Two-person bench-type, padded seat and back rest
- ⑱ Foot gong
- ⑲ Bulkhead mounted chain locked hand brake
- ⑳ Reversible, swiveling, removable, wood trolley pole with slider shoe
- ㉑ Easy access equipment inspection covers and doors
- ㉒ Single, automatically controlled "quiet idle", two-speed blower
- ㉓ Air operated, frame mounted sanders at all wheels
- ㉔ Air horn
- ㉕ Multiple cylinder air brake system with standard railroad brake components
- ㉖ Grease-lubricated, anti-friction, roller-type motor axle bearings. Grease hoses furnished for side lubrication
- ㉗ Steel-tyred wheels
- ㉘ Battery charging alternator with integral rectifier and regulator
- ㉙ Rounded end frames

General Electric Mining Locomotive Safety And Productivity

If you're operating a vintage mining locomotive, you can improve performance, reliability and safety by upgrading its motor control system with General Electric's cost-effective modernization kit.

The GE power system includes contactors, interlocks, relays, and the latest in electrical components. All are integrated according to control circuits designed for locomotives ranging from 6 to 15 tons. The system has been applied at greater weights.

Designed as a straight parallel system, the GE power system kit offers significant advantages over older series-parallel designs.

Consider performance. GE's parallel design provides improved starting because wheels recover better in the event of wheel slippages. And, a greater number of acceleration and deceleration steps smooths locomotive handling.

Heart of the GE power system is a controller unit which provides eight steps of acceleration on 6 to 8-ton locomotives and 10 steps on locomotives over 9 tons. The result is smoother handling and less chance of pulled drawbars due to jackrabbit starts.

In the deceleration mode, the controller provides six steps on 6 to 8-ton locomotives and seven steps on locomotives over 9 tons. GE's dynamic braking system results in smoother handling and reduced brake and wheel wear. It also qualifies as alternate braking equipment under existing mine safety requirements. Since this system cannot stop the unit, a stopping braking system must be provided.

Reflect on safety. GE's control system provides greater operator safety by replacing older high voltage systems with a 37-volt system.

Other safety features include failsafe operation which removes power in case of operator disability. Air brake can be provided on locomotives with an air system. Other systems would have to be reviewed. Should locomotive trolley contact be lost, dynamic braking remains operable and lighting is maintained.

Although a 34 or 37-volt system requires batteries and charging equipment, it is not included in the kit but can be obtained from GE or local sources.

Opinions vary as to the optimal battery charging system and so the decision is left to the individual operator. GE, however, believes that a motor-alternator system offers the best overall performance and will provide information upon request.

Think about maintenance. Chances are your present system is becoming more and more costly to maintain because of increased downtime and delivery hold-ups due to hard-to-obtain parts. Standardize your control system with up-to-date GE controls and we'll be more responsive to your equipment replacement needs at less overall expense.

GE's power system is also simpler to maintain and repair than your present system because it contains fewer contactors, relays and other components. Power overload protection is provided by easy-to-reset overload relays which replace power fuses.

When properly integrated into the locomotive, GE control system components will help reduce out-of-service time, increase parts availability, and improve mine locomotive performance.

To process your modernization request, we'll need to know the serial number of your existing GE locomotive. For other models, we'll need to know the manufacturer, model weight, wheel diameter, gear ratio, motor horsepower, and voltage.

To order your modernization kit, write to C.W. Iverson, General Electric Co., Transportation Systems Business Division, 2901 East Lake Road, Erie, PA 16531.

General Electric Modernization Kit Components

Description	Part Number	Quantity 6 to 8 tons	Quantity 9 tons and up Contactor Reversing	Quantity 9 tons and up Air Reversing
Contactor	17CM53K9	10	19	12
Interlock	17AF14H3	4	14	6
Contactor	17CM57N6A	1	1	1
Interlock	17AF14K4	1	1	1
Overload Relay	17LS16K (Single Motor)	2	2	2
Overload Relay	17LSK6K (Trolley Circuit)	1	1	1
Time Delay Relay	41A264547P17	1	1	1
Time Delay Relay	41A264547P3 (250V)	1	1	1
Blocking Diode	41A562325G1	2	2	2
Safety Foot Switch	41A300302G2	1	1	1
Controller	17KC62C1		1	1
Accelerating Resistor		1	1	1
Reverser	17DP24A3			1
Controller	41B540634P1	1		
Switch	17GC7E2	1		

Other miscellaneous components shown on the wiring diagram will be specified upon request.

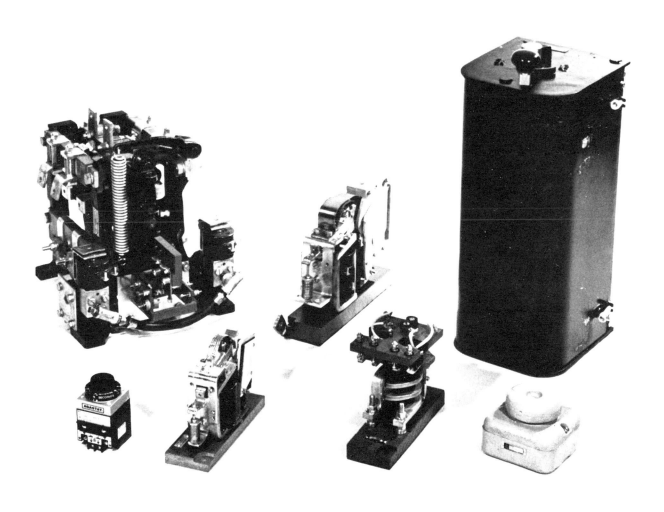

Power Circuits

Mine Locomotives, 6 And 8-Ton

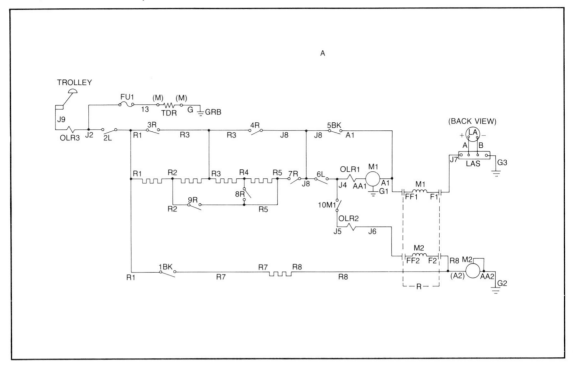

Mine Locomotives, 9-Ton And Up With Contactor Reversing

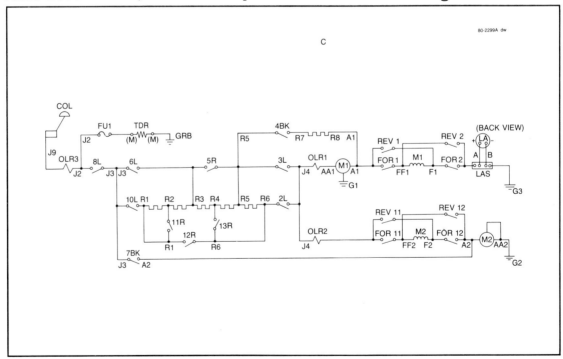

Control Circuits
Mine Locomotives, 6 And 8-Ton

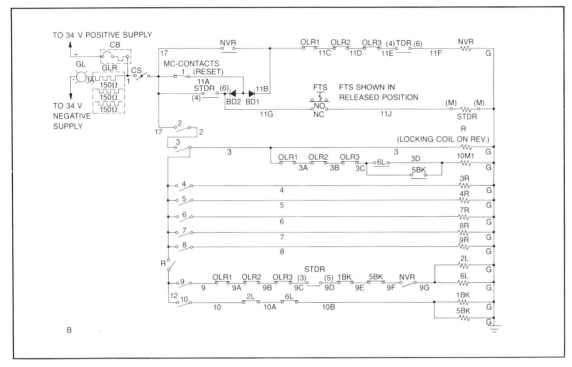

Mine Locomotives, 9-Ton And Up With Contactor Reversing

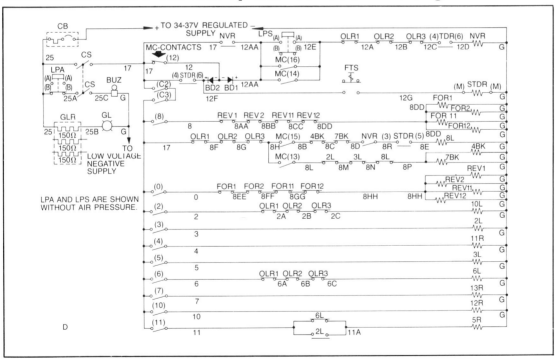

GENERAL ELECTRIC COMPANY

LOCOMOTIVE MILESTONE EVENTS

1880

The first experimental electric lomomotive was operated by Edison at Menlo Park, New Jersey, using many of his new inventions for drive and braking systems. Eventually a total of some 1300 patents were issued to Thomas A. Edison.

1887

The first mine locomotive, having 40hp, was produced for the Lykens Valley Colliery of the Pennsylvania R.R.

1888

Van Depoele developed carbon brushes for railway motors which had much to do with the success of the direct current DC motor.

The Thomson-Houston Co. installed the first industrial electric locomotive at the Tremont & Suffolk Mills in Lowell, MA.

1892

The General Electric Company was formed from the Edison General Electric Co. and the Thomson-Houston Co. with all their respective subsidiaries.

The Baltimore & Ohio Railroad became the first steam railroad in U.S.A. to use electric locomotives and power equipment. Three heavy duty 96 ton 1440 hp locomotives (4—360 hp gearless motors) to eliminate smoke-filled tunnel hazards.

A complete line of mine locomotives was offered from 2 to 11 tons, with motor capacities from 15 to 150 hp, with speeds ranging from 6 to 10 mph.

1893

The GE Transportation Department operated a complete electric elevated railroad, the Intramural Railway, hauling passengers around the fairgrounds of the Chicago Columbian Exposition.

1895

The traction motor was developed into its present form and appearance.

1898

The first GE export electric locomotive was shipped to the London Central Underground, (for the London, England subway system).

1904

GE electrified New York Central's Grand Central Station in New York City.

New York Central ordered thirty 113 ton, 1700 hp, 42 mph, bi-polar completely gearless electric locomotives for 600 volts Third Rail use in Grand Central Station. Using Sprague's multiple-unit system, 2 locomotives coupled together were able to haul the heaviest loads yet handled. All of these locomotives operated over 50 years before retirement.

1905

First order placed for terminus-to-terminus railroad electrification from Camden to Atlantic City, New Jersey.

1906

A prototype, self-propelled gas-electric railcar, Delaware & Hudson Railway number 1000, placed in service between Schenectady and Saratoga, New York. It had two 200 hp, motors supplied by a 120 kw generator of 6000 volts, driven by a powerful gasoline engine, the first V8 engine in the U.S.A.

1907

The first motor-generator type electric locomotive placed in service on the Paul Smith Railroad in the New York Adirondack Mountains. The original locomotive had rectifier tubes later replaced by the motor generator.

An electric cable gathering reel is developed for mine locomotives.

1909

GE electrified the Great Northern Railway's Cascade Tunnel, 6600 volts at 25 Hz. to operate 1500 hp motors.

1911

GE's Pennsylvania property became the Erie Works.

Manufacturing started on gas-electric car equipment at Erie; eventually 89 units were built.

1913

GE built its first commercial gasoline-electric locomotive, number 100, with 2-GE V-type 175 hp gasoline engines, for the Minneapolis, St. Paul, Rochester and Dubuque Electric Traction Co., a predecessor of the Minneapolis, Northfield & Southern Railway. This locomotive, although gas-electric, was eventually the forerunner of all diesel-electric locomotives. Today it is preserved as a working historical exhibit.

1914

Dr. Hermann Lemp patented the famous Lemp Control Circuit for matching electro-magnetic characteristics of traction generation to the internal combustion engine horsepower curve. All diesel-electric locomotive control systems incorporate this basic principle.

1915

GE installed the first U.S. 3000 volts DC mainline electrification of the Chicago, Milwaukee & St. Paul Railroad, for their 440 mile route, from Harlowton, Montana to Avery, Idaho. Fourty-two 260 to 280 ton electric locomotives, using regenerative braking for steep grades, were delivered.

1918

The first diesel engines for railroad use were built at the Erie Works, and installed on 3 experimental steeple-cab diesel-electric locomotives, which were equipped with Lemp controls. These were sold to: The Jay Street Connecting Railroad of New York City, the City of Baltimore, and the U.S. Army.

GE built the first 3000 volts DC, 260 ton bi-polar, gearless electric passenger locomotives for the Chicago, Milwaukee & St. Paul Railroad, and were tested and exhibited at the Erie Works.

1920

Development of a 50 ton frameless truck locomotive.

A complete line of battery mine locomotives offered from 1-1/2 to 12 tons.

The great tug-of-war at Kent, Washington between Chicago, Milwaukee & St. Paul Railroad number 10253, a 260 ton bi-polar electric locomotive against 2 steam locomotives (one articulated). Although the steam locomotives got a head start, the GE electric traction motors soon retarded the steam locomotives lead, reversed the lead, and succeeded in out-pulling the steam locomotives, winning the event.

1924

The World's first diesel-electric switcher locomotive, a B-B type, 60 ton unit with an Ingersoll-Rand 300 hp engine, powered by Lemp controls was placed in service December 17, 1924. Commercially successful, it was purchased by the Central Railroad of New Jersey, road number 1000 and retired from service in 1957.

1927

GE first demonstrated radio communication from the caboose to the locomotive of a one-and-a-quarter mile long freight train.

1928

The first diesel-electric passenger locomotive was placed in service on the New York Central Railroad. A 2-D-2 type, with a McIntosh & Seymour engine rated at 900 hp, it was of the air injection type. Electrical components were by GE and it was built by the American Locomotive Company.

1934

Delivery began to the Pennsylvania Railroad of the famous GG1, 100 mph high-speed electric locomotives.

The "Pioneer Zephyr", a 3-car experimental-type, lightweight stainless-steel streamlined diesel-electric passenger train was delivered to the Chicago, Burlington & Quincy Railroad, Built by Budd, it had an Electro-Motive Winton diesel engine with GE electrical components.

1935

The first non-articulated high-speed diesel-electric passenger locomotive was built by GE, to the specifications of Electro-Motive, and was placed in service on the Baltimore & Ohio Railroad in August. It was a box-style unit of the A1A A1A type, developing 1800 hp with 2 Electro-Motive Winton 900 hp diesel engines.

1936

GE built the first single engine 2000 hp diesel-electric locomotive. Placed in service on the Illinois Central Railroad, it had a Busch-Sulzer diesel engine.

1937

Two steam-turbine electric locomotives were built for the Union Pacific Railroad, however, after extensive testing were scrapped.

1938

A complete line of standard diesel-electric switcher locomotives was offered in these models:

	25 tons	45 tons	65 tons	80 tons
(in metric measurement)	23 tonnes	41 tonnes	59 tonnes	73 tonnes

1939

A diesel-electric locomotive was built for the Pike's Peak Cog Railway, which has the highest altitude at which diesel-electric locomotives operate.

1940

Announcement was made of joint-merchandising a line of diesel-electric locomotives by both General Electric and American Locomotive (ALCO). Generally the larger switchers and mainline locomotives were built by ALCO, with GE electrical components; generators, motors and controls. Smaller diesel-electric switcher locomotives were built by GE.

1944

The use of sealed lubrication on traction motor armature bearings was pioneered by GE.

ALCO-GE started use of welded truck frame assemblies on high speed locomotives.

1945

GE initiated a program to develop and build a gas-turbine electric locomotive.

The first Fairbanks-Morse passenger and freight diesel-electric locomotives were built by GE. These were 3 units, each 2000 hp of the A1A-A1A type built at the Erie Works for the Union Pacific Railroad.

1946

GE built 20 large electric locomotives, of broad gauge for the U.S.S.R. (Russia). Nicknamed "Little Joes", they were not shipped to the U.S.S.R., but were sold as follows: 12 to the Milwaukee Road, 3 to the South Shore Line and 5 to Brazil's Paulista Railway.

1947

The largest electric locomotives of the time, two 11000 volts B-D+D-B types built for the Great Northern Railway.

1948

The first U.S. gas turbine electric locomotive, a B-B-B-B type (four, four wheel trucks) rated at 4500 hp, was built at Erie Works. The double ended cab unit was the first tested on the Pennsylvania Railroad, then the Nickel Plate Road, and finally over a year on the Union Pacific Railroad, handling heavy freight trains.

1950

GE received the first order for gas-turbine electric locomotives, ten 4500 hp units for the Union Pacific Railroad.

1951

For the first time in America, Mica-mat insulation was used in traction motors.

1952

GE initiated its mainline diesel-electric locomotive development program.

Second order received for gas-turbine electric locomotives, fifteen 4500 hp units for the Union Pacific Railroad.

1953

The first U.S. diesel-electric mine locomotive was delivered.

1954

A new 4 unit diesel-electric locomotive "rolling lab", in Erie Railroad colors, as placed in mainline freight service demonstrate and evaluate diesel-electric locomotive components.

1955

A laboratory, to further development of locomotive diesel engines, was installed at the Erie Works.

Ten of the first high speed 4000 hp rectifier-equipped electric passenger locomotives were shipped to New York, New Haven & Hartford Railroad.

1956

The Universal Line, a new standard line of mainline diesel-electric locomotives was offered for World-wide service.

1958

The World's largest internally-powered locomotive, an 8500 hp gas-turbine electric locomotive was shipped to the Union Pacific Railroad, later re-rated to 10000 hp.

1959

The first two 2400 hp prototype diesel-electric locomotives were built for testing on U.S. Railroads.

Shipment began on the largest order ever placed with the GE Locomotive & Car Equipment Department, 115 U181 Model diesel-electric locomotives for the South African Railways.

The Universal Line was expanded to include the U6B Model.

New comprehensive adhesion loss detection and correction systems were incorporated in locomotives.

1960

Introduction of the U25B Model, GE's 2500 hp 4-axle powered diesel-electric locomotive for high-speed freight service on U.S. Railroads.

Shipment of the first of 66 4400 hp rectifier-type electric freight locomotives for mainline service, to the Pennsylvania Railroad which was the largest order received in the U.S. for rectifier-type electric locomotives.

1961

GE increased the horsepower ratings of its 8 and 12 cylinder diesel-electric locomotives by about 10%.

Shipment of the first U25B Model diesel-electric locomotives to U.S. Railroads.
The U25B Model received a low front hood.

GE designed the first transistorized rapid transit control equipment, and the automated rapid transit equipment was demonstrated at the Erie Works test track.

1962

Shipment of the first 4400 hp silicon rectifier-type electric locomotives to the Pennsylvania Railroad, which is the first application of solid-state rectifiers to locomotive service.

GE transferred test and assembly facilities of the FDL diesel engine to the Erie Plant.

Introduction of new industrial diesel-electric locomotives designed for steel mill service.

A radio remote-controlled industrial locomotive tested.

GE supplied motors and controls for the monorail trains at the Seattle World's Fair.

1963

Introduction of the U25C Model, a 2500 hp six axle powered mainline diesel-electric locomotive.

GE designed the U50 Model, a 5000 hp, eight axle powered mainline diesel-electric locomotive, powered by 2 completely independent 16 cylinder diesel engine drive systems.

1964

The diesel-electric switcher line expanded to offer these models:

	100 tons	110 tons
(in metric measurement)	91 tonnes	100 tonnes

1965

The first mainline diesel-electric locomotive with GE alternators instead of DC traction generators built.

1967

Introduction of the U30 Model, a 3000 hp mainline diesel-electric locomotive, featuring new electrical systems with high performance alternator, solid state circuitry and GE 752 motors to optimize performance of the units transmission systems.

First application made of slow speed regulation equipment to a diesel-electric locomotive, for unit train operations, went into service on the Pennsylvania Railroad.

1968

Introduction of the U33C Model, a 3300 hp mainline diesel-electric locomotive which had 70 mph gearing and 90,000 pounds of continuous tractive effort, and is capable of performing both high-speed and high-tonnage services.

Delivery was made of the first fully-automated 5000 hp rectifier-type electric locomotives, using commercial frequency, to the Muskingum Electric Railroad.

1969

GE built the first fully-automated AC coke-quencher locomotive system for Bethlehem Steel.

First high-speed fully-automated position stop rapid transit cars placed in service on the Port Authority Transit Corporation Lines.

1970

The U34CH Model diesel-electric commuter locomotive entered suburban service on the Erie Lackawanna Railway. The diesel engine drives both traction and auxiliary alternators, the latter eliminating the need for a separate engine-alternator set for train power.

1971

GE developed MATE, a non-powered unit providing Motors for Additional Tractive Effort to haul heavy trains in conjunction with a locomotive, generally for heavy switching operations.

1972

Delivery made of the World's first 50000 volts, 6000 hp thyristor controlled electric locomotives for automated operations, on the Black Mesa & Lake Powell Railroad, a coal-hauling line for an electric power generating plant.

1973

GE introduced the Constant Horsepower Excitation Control. In establishing a unique marriage between the turbocharger and governor-controlled engine, the system regulated the amount of fuel going to the diesel engine, reducing smoke pollution, and improving fuel economy.

1974

Introduction of a standard line of diesel-electric switcher locomotives offered in these models:
in metric measurement (second line)

65-84 tons/600hp	85-100 tons/600hp	115 tons/800hp	125-144 tons/1100hp
59-76 tonnes/609CV	77-100 tonnes/609 CV	105 tonnes/810 CV	113-131 tonnes/1116CV

1975

The multi-voltage E60C Model electric locomotive built for AMTRAK, capable of 120 mph, with a continuous rating of 6000 hp, it has a 10000 hp shorttime acceleration rating for a few minutes.

1977

Introduction of the Dash 7 New Series mainline diesel-electric locomotives, which greatly improved reliability, availability, maintainability and incorporating design changes from 15 major railroad operators combining over 60 other improvements.

1981

SENTRY Adhesion Control System introduced for improved adhesion to allow for full utilization of the higher continuous current rating of the new 752*AP Traction Motor. *752 is a GE Trademark.

1982

39 6000 hp 70mph passenger/freight double-ended electric locomotives built for Mexico. These were equipped with Solid-State circuitry, thyristor controls and vacuum breakers.

1984

First Dash 8 locomotive tested.

GE built the World's most technologically advanced locomotive fabrication plant. A $100 Million facility, about 50 miles South of Erie, at Grove City, PA., it is capable of producing up to fifteen engines weekly — for final completion at the Erie Plant.

1987

Ultra-efficient Line of Dash 8 diesel-electric locomotives introduced. These comprise the Dash 8-40C, Dash 8-40B, Dash 8-32C and Dash 8-32B Models.

GE sustains its marketing and sales dominance of the U.S. diesel-electric locomotive market.

1988

Canadian National Railways ordered 30 Dash 8-40C 4000 hp locomotives with the Safety Deluxe Canadian Cabs.

GE Canada bought from Bombardier the Montreal Locomotive Works, once owned by ALCO.

DASH 8 DIESEL-ELECTRIC LOCOMOTIVES

OPERATOR'S MANUAL

© 1987 GENERAL ELECTRIC COMPANY

LOCATION OF APPARATUS

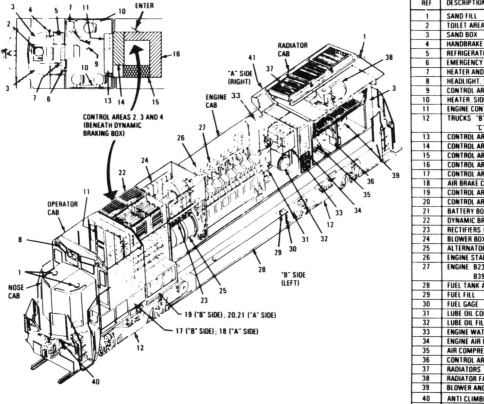

REF	DESCRIPTION
1	SAND FILL
2	TOILET AREA
3	SAND BOX
4	HANDBRAKE
5	REFRIGERATOR OR COOLER
6	EMERGENCY BRAKE VALVE
7	HEATER AND DEFROSTER
8	HEADLIGHT, SIGNAL AND NUMBER LIGHT BOX
9	CONTROL AREA #5 (CONTROL CONSOLE)
10	HEATER, SIDE STRIP
11	ENGINE CONTROL PANEL
12	TRUCKS "B" 2 AXLES PER TRUCK / "C" 3 AXLES PER TRUCK
13	CONTROL AREA #1
14	CONTROL AREA #2 — LOCATED IN AUXILIARY CAB
15	CONTROL AREA #3 — LOCATED IN AUXILIARY CAB
16	CONTROL AREA #4
17	CONTROL AREA #6
18	AIR BRAKE COMPARTMENT
19	CONTROL AREA #7
20	CONTROL AREA #8
21	BATTERY BOX
22	DYNAMIC BRAKING BOX
23	RECTIFIERS (PROPULSION)
24	BLOWER BOX AND AIR FILTERS
25	ALTERNATORS (MAIN AND AUXILIARY)
26	ENGINE START STATION
27	ENGINE B23, B32, C32 (12 CYLINDER, 7FDL12) B39, C39 (16 CYLINDER, 7FDL16)
28	FUEL TANK AND RETENTION TANK
29	FUEL FILL
30	FUEL GAGE
31	LUBE OIL COOLER
32	LUBE OIL FILTER
33	ENGINE WATER TANK AND WATER CONTROL VALVE
34	ENGINE AIR FILTER COMPARTMENT
35	AIR COMPRESSOR (MOTOR DRIVEN)
36	CONTROL AREA #9
37	RADIATORS
38	RADIATOR FAN
39	BLOWER AND AIR FILTERS (NO 2 END)
40	ANTI-CLIMBER (OPTIONAL)
41	TOOL HOLDER BOARD

CONTENTS

	Page
LOCATION OF APPARATUS	218
GENERAL DATA	219
OPERATING CONTROLS	
INTRODUCTION	220
DEVICES ON CONTROL CONSOLE	220
DEVICES ON ENGINE CONTROL PANEL	222
OTHER OPERATOR CAB CONTROLS	224
DIAGNOSTIC DISPLAY PANEL (DID)	
GENERAL INFORMATION	226
THE DISPLAY	226
USING THE DISPLAY	226
MESSAGES AT LOCOMOTIVE START-UP	226
OPERATING MODES IN LEVEL 1	227
EXAMPLE - LEVEL 1 OPERATION	229
LIST OF SUMMARY MESSAGES	232
AIR BRAKE EQUIPMENT	
AIR EQUIPMENT ON THE CONTROL CONSOLE	232
AIR BRAKE EQUIPMENT IN AIR BRAKE COMPARTMENT	233
AIR COMPRESSOR SAFETY VALVES	233
CUT-OUT COCKS	233
ADJUSTING VALVES	234
OTHER EQUIPMENT	
CONTROL COMPARTMENT EQUIPMENT	236
EQUIPMENT BLOWERS AND RADIATOR FAN	239
AIR COMPRESSOR	240
DIESEL ENGINE CONTROL GOVERNOR	241
MISCELLANEOUS EQUIPMENT	241
GAGES AND MEASURING DEVICES	242
PRESSURE AND TEMPERATURE GAGES	242
OTHER GAGES	242
DRAINING COOLING WATER SYSTEM	243
ALARMS, SAFEGUARDS, POWER DERATIONS AND SHUTDOWNS	243
BARRING-OVER SWITCH	243
EMERGENCY SANDING	244
ENGINE AIR FILTER PRESSURE SWITCH	244
GROUND CUT-OUT SWITCHES	244
MOTOR CUT-OUT SWITCHES	244
OIL AND WATER TEMPERATURE AND PRESSURE	244
OVERSPEED - ENGINE SHUTDOWN	245
OVERSPEED - LOCOMOTIVE	245
PCS SWITCH OPERATION	245
POWER LIMIT SWITCH	245
SAFETY CONTROL FOOT PEDAL	245
WHEELSLIP	245
PREPARATION FOR OPERATION	
BEFORE BOARDING LOCOMOTIVE	246
AFTER BOARDING LOCOMOTIVE	246
STARTING ENGINE	246
BEFORE MOVING LOCOMOTIVE	247
FASTER AIR PUMPING	247
COLD WEATHER ENGINE STARTING/WARM UP	247
OPERATING PROCEDURES	
MOVING A TRAIN	247
STOPPING A TRAIN	247
REVERSING LOCOMOTIVE	248
PASSING THROUGH WATER	248
PASSING OVER RAILROAD CROSSINGS	248
STOPPING ENGINE	248
BEFORE LEAVING LOCOMOTIVE	248
SAFETY CONTROLS	248
DYNAMIC BRAKE OPERATION	
APPLYING DYNAMIC BRAKES	249
USE OF AIR BRAKES DURING DYNAMIC BRAKING	249
RELEASE OF DYNAMIC BRAKING	249
MULTIPLE-UNIT OPERATION	
OPERATING AS A LEADING UNIT	249
OPERATING AS A TRAILING UNIT	249
CHANGING OPERATING ENDS	250
TO OPERATE WITH OTHER TYPES OF UNITS	250
BRAKE PIPE LEAKAGE TEST	250
DEAD HEADING (DEAD-IN-TRAIN)	250

	Dash 8-32B (B32-8)	Dash 8-32C (C32-8)	Dash 8-40B (B39-8)	Dash 8-40C (C39-8)
Operating Cab and Controls	General Purpose	General Purpose	General Purpose	General Purpose
Wheel Arrangement	B-B	C-C	B-B	C-C
Engine Data				
Horsepower - Traction	3150	3150	4000 (3900)	4000 (3900)
Number of Cylinders	12	12	16	16
Model	GE FDL12	GE FDL12	GE FDL16	GE FDL16
Bore and Stroke (in.)	9 x 10-1/2	9 x 10-1/2	9 x 10-1/2	9 x 10-1/2
RPM (max.)	1050	1050	1050	1050
Compression Ratio	12.7:1	12.7:1	12.7:1	12.7:1
Cycle	4	4	4	4
Turbocharged	Yes	Yes	Yes	Yes
Engine Cooling Fan	1	1	1	1
Engine Cooling Fan Drive	A-C Motor	A-C Motor	A-C Motor	A-C Motor
Traction Equipment				
Traction and Auxiliary Alternator	GMG 186	GMG 187	GMG 186	GMG 187
Alternator Blower	1	1	1	1
Traction Motor	4-GE752	6-GE752	4-GE752	6-GE752
Traction Motor Blowers	2	2	2	2
Blower Drives	A-C Motor	A-C Motor	A-C Motor	A-C Motor
Air Brake Schedule	26L	26L	26L	26L
Major Dimensions				
Length	63 ft., 7 in.	67 ft., 11 in.	66 ft., 4 in.	70 ft., 8 in.
Height	14 ft., 11-1/2 in.	15 ft., 4-1/2 in.	14 ft., 11-1/2 in.	15 ft., 4-1/2 in.
Width	10 ft., 2-3/4 in.	10 ft., 2-3/4 in.	10 ft., 2-3/4 in.	10 ft., 2-3/4 in.
Bolster Centers	36 ft., 7 in.	40 ft., 7 in.	39 ft., 4 in.	43 ft., 4 in.
Truck Wheel Base	9 ft., 0 in.	13 ft., 7 in.	9 ft., 0 in.	13 ft., 7 in.
Minimum Track Curvature (rad. and deg.):				
For Single Unit	150 ft./39°	273 ft./21°	150 ft./39°	273 ft./21°
For MU	195 ft./29°	273 ft./21°	195 ft./29°	273 ft./21°
Driving Wheel Diameter (in.)	40	40	40	40 (42 optional)

	Dash 8-32B (B32-8)	Dash 8-32C (C32-8)	Dash 8-40B (B39-8)	Dash 8-40C (C39-8)
Weight				
Per Axle (pounds minimum and maximum)	67,250/70,000	57,717/70,000	68,500/72,000	61,167/70,000
Total (pounds minimum and maximum)	269,000/280,000	346,300/420,000	274,000/288,000	367,000/420,000
Tractive Effort (pounds)				
Starting at 25% Adhesion for Minimum and Maximum Weight	67,250/70,000	86,575/105,000	68,500/72,000	91,750/105,000
Cont. Tractive Effort and Speed (mph)				
For Smaller Pinion (83/20)	70,140 @ 13.9	108,360 @ 8.2	68,100 @ 18.3	106,790 @ 10.9
For Larger Pinion (81/22)	62,230 @ 15.7	96,140 @ 9.2	60,420 @ 20.6	94,740 @ 12.3
Gear Ratio and Max. Speed (mph)				
Smaller Pinion (83/20)	70	70	70	70
Larger Pinion (81/22)	79	79	79	79
Supplies				
Fuel Tank (gal.)	3150	3900	3900	5000
Coolant (gal.)	350	350	380	380
Lube Oil (gal.)	260	260	360	360
Sand (cu. ft.)	48	48	48	48
Compressor, Air				
Compressor Drive	A-C Motor	A-C Motor	A-C Motor	A-C Motor
Maximum Delivery CFM	296	296	296	296
Type of Cooling	Air	Air	Air	Air
Draft Gear	NC390	NC390	NC390	NC390
Air Filtering Devices				
Primary	Vortex Self-Cleaning	Vortex Self-Cleaning	Vortex Self-Cleaning	Vortex Self-Cleaning
Secondary Engine Air Intake	AAF	AAF	AAF	AAF
Engine Room Pressurized	Yes	Yes	Yes	Yes
Main Generator Pressurized	Yes	Yes	Yes	Yes

OPERATING CONTROLS

INTRODUCTION

All of the operating devices, manual and visual, normally used by the operator during locomotive operation are located near the operator's position. Most of these devices are located either on the control console or on the Engine Control panel.

NOTE: *Customer equipment requirements often differ from one railroad to another. Therefore, physical locations and appearance of some devices illustrated in this manual may not agree entirely with the equipment furnished to any particular railroad.*

DEVICES ON CONTROL CONSOLE (Fig. 2)

The following operating devices are located on the control console:

Master Controller

The Master Controller is a Set-Up switch used by the operator to control the locomotive during operation. It is equipped with a Throttle handle, Dynamic Braking handle and Reverse handle.

Reverse Handle

The Reverse handle, the bottom of the three handles, is used to determine the direction of locomotive travel. It has positions REVERSE, OFF and FORWARD. The handle is removable only when the Throttle handle is in IDLE position and Braking handle is in OFF.

Throttle Handle

The Throttle handle is the middle handle. It has a SHUTDOWN, IDLE and eight major positions or notches for power.

The SHUTDOWN position is located to the right of IDLE and is used in an emergency to shut down all engines of a multiple-unit consist from the operator's position of the controlling unit. Pull out axially on Throttle handle and move the handle to the right to put it into SHUTDOWN.

To increase motoring power, the handle is moved clockwise toward the operator.

Braking Handle (Dynamic Braking)

The Braking handle is above the Throttle handle and has OFF and SET-UP positions and a notchless BRAKING sector.

In the OFF position, nearest the operator, dynamic braking is shut off. The SET-UP position establishes dynamic braking circuits. Movement beyond this position into the BRAKING sector (counterclockwise away from the operator) increases braking effort.

Interlocking Between Handles

Interlocking between the handles of the Master Controller is provided as follows:

1. The Reverse handle must be inserted before the Throttle handle can be moved out of IDLE position for power or emergency shutdown.

FIG. 2. OPERATOR'S CONTROL CONSOLE.

REF	DESCRIPTION
1	MASTER CONTROLLER
2	LOAD AMMETER
3	ENGINE RUN BREAKER
4	GENERATOR FIELD CIRCUIT BREAKER
5	CONTROL CIRCUIT BREAKER
6	POWER LIMIT SWITCH
7	DYNAMIC BRAKING CONTROL CIRCUIT BREAKER
8	GAGE LIGHT DIMMER KNOB (ON SIDE OF CONSOLE)
9	FRONT HEADLIGHT SWITCH
10	TRAINLINE GROUND RESET BUTTON
11	CALL BUTTON
12	REAR HEADLIGHT SWITCH
13	HUMP CONTROL (OPTIONAL)
14	STEP LIGHT SWITCH
15	GAGE LIGHTS SWITCH
16	WHEELSLIP LIGHT
17	PCS OPEN LIGHT
18	DYNAMIC BRAKE WARNING LIGHT
19	SAND LIGHT
20	LEAD AXLE SAND SWITCH
21	SAND SWITCH
22	BELL VALVE
23	HORN VALVE
24	RADIO LOCATION (OPTIONAL)
25	AIR GAGES
26	BRAKE PIPE FLOW INDICATOR (OPTIONAL)
27	AUTOMATIC BRAKE VALVE HANDLE
28	INDEPENDENT BRAKE VALVE HANDLE
29	BRAKE PIPE CUT OUT PILOT VALVE
30	MU2A OR DUAL PORTED CUT OUT COCK
31	BRAKE PIPE REGULATING VALVE
32	SAFETY CONTROL PEDAL (OPTIONAL)
33	SELECT A POWER, FUEL SAVER (OPTIONAL)

FIG. 2. OPERATOR'S CONTROL CONSOLE.

OPERATING CONTROLS

2. The Reverse handle can be moved into FORWARD or REVERSE only when the Throttle handle is in IDLE position and the Braking handle is in OFF position.

3. The Reverse handle cannot be moved out of FORWARD or REVERSE position when either the Throttle handle is advanced beyond IDLE or the Braking handle is advanced beyond OFF.

4. The Braking handle must be in OFF position before the Throttle handle can be moved out of IDLE position, except for emergency shutdown.

5. The Throttle handle must be in IDLE and the Reverse handle in FORWARD or REVERSE before the Braking handle can be moved.

6. The Reverse handle can be removed only when the Reverse handle is CENTERED, Braking handle is in OFF and the Throttle handle is in IDLE.

Operation

To manipulate the controller operating handles during locomotive operation, proceed as follows:

Lead or Single-Unit Operation

Operating Handle Set-Up (Reverse handle removed):

1. Braking handle in OFF.
2. Throttle handle in IDLE.

> **CAUTION:** *Finding the Braking handle away from OFF or the Throttle handle away from IDLE with the Reverse handle removed indicates that interlocking between handles requires repair or adjustment. Do not attempt to operate.*

3. Insert the Reverse handle.
4. Set Reverse handle for the desired direction of operation.

Operating in Power Mode:

1. Braking handle remains in OFF
2. Move Reverse handle to desired position
3. Move the Throttle handle to the desired notch.

Operation in Dynamic Brake Mode:

1. Throttle handle returned to IDLE
2. Move the Braking handle to SET-UP; pause, then advance as desired.

Operation as Trail Unit:

1. Braking handle in OFF
2. Throttle handle in IDLE
3. Reverse handle centered and removed.

For Emergency Multiple-Unit Shutdown:

In the controlling unit, pull out axially on the Throttle handle, and move it beyond IDLE to SHUTDOWN.

> **NOTE:** *In a Trail unit, the Reverse handle must be inserted to release the Throttle handle before it can be moved to SHUTDOWN.*

Load Ammeter

This meter shows the average current going to each of the traction motors. Motoring is shown to the right of 12 o'clock and has two bands. The Green band is in the continuous rating of the motors and the Red band is the short-time rating.

> **CAUTION:** *The control system on this locomotive automatically limits the time in short-time rating to protect the equipment. If other locomotives in the consist do not have this feature, observe the time limit on the meter when in the short-time rating to prevent equipment damage.*

Dynamic Braking is shown to the left of 12 o'clock and has two bands. The Yellow band is the continuous rating of the motors and the Red band is overload. If the meter goes into the Red band, reduce the Braking handle position until the meter goes back into the Yellow band.

Engine Run Circuit Breaker

The Engine Run breaker controls engine speed. It must be ON to control engine speed of the Lead locomotive and all units of a consist. On Trail units, the breaker is in the OFF position.

Generator Field Circuit Breaker

The Generator Field circuit breaker is ON whenever the locomotive is powered and operating as a Lead unit. The breaker may be turned off to keep the main generator de-energized when it is necessary to run the engine at speeds higher than IDLE. On Trail locomotives, it is in the OFF position.

Control Circuit Breaker

The Control breaker must be ON to run the fuel pump, start the engine and provide power to other circuits, including the auxiliaries. In MU operation, this breaker must be ON on the Lead unit only.

Power Limit Switch

> **NOTE:** *This switch may be eliminated as a customer option and may therefore not appear on all locomotives.*

This switch has two positions, NORMAL and NOTCH 7.

When the Leading unit is slipping excessively, the Power Limit switch can be moved to NOTCH 7 to reduce power while the Trailing units are operating at full power. This will reduce the tractive effort on the Leading unit and will usually improve the ability of the locomotive to hold the rail under bad rail conditions.

Also see ALARMS, SAFEGUARDS, POWER DERATIONS AND SHUTDOWNS section of this manual.

> **NOTE:** *Unless directed otherwise by railroad rules, make sure the Power Limit switch is in NORMAL position on ALL units when boarding the train.*

Dynamic Braking Control Breaker

> **NOTE:** *Dynamic Brake is provided as optional equipment.*

The Dynamic Braking Control breaker is used to control the dynamic braking of the locomotive. In MU operation, this breaker must be ON on the Lead unit only to control the dynamic braking of other units in the consist.

OPERATING CONTROLS

Gage Light Dimmer Knob

The dimmer knob is located on the right side of the console. It is used to brighten and dim the console gage lights.

Front Headlight Switch

This switch controls the operation of the front headlight and has four positions; OFF, DIM, MED and BRIGHT.

Trainline Ground Reset Button (Optional)

Resets the Ground Relay on locomotives which are equipped for trainline ground reset. New Series-8 locomotives may transmit the Reset signal, but do not respond to it.

Call Button

The Call button is used to sound the alarm bell in all locomotive units. This button can be used to test the alarm bell when boarding the locomotive.

Rear Headlight Switch

This switch controls the operation of the front headlight and has four positions; OFF, DIM, MED and BRIGHT.

Power Reduction (Hump) Control (Optional)

Allows operator to precisely control locomotive power outputs. The Hump Control toggle switch has three positions; OFF, LOCAL and TRAINLINE. This toggle switch is set for the desired operation, then the Throttle handle is advanced to the desired notch. The Hump Control potentiometer is then positioned between MIN and MAX to meet horsepower requirement.

Step Light Switch

Turns on all four corner step lights.

Gage Light Switch

Turns on the operator console gage lights.

WHEELSLIP Light

This light, accompanied by an optional buzzer, indicates that the wheels on some locomotives in the consist are slipping. This is a trainlined indication.

PCS OPEN Light

Indicates a Penalty or Emergency air brake application has occurred and power has been limited or removed.

DYNAMIC BRAKE WARNING Light

This light, accompanied by an optional buzzer, indicates that a locomotive in the consist is experiencing excessive dynamic braking current. Reduce the Braking handle position until this light goes out.

SAND Light

Indicates that sanding is taking place either manually or as a result of wheelslip. During a wheelslip, sanding and the SAND light will automatically turn on, then off.

Lead Axle Sand Switch

Sand will be applied to rail in front of the leading axle, depending on locomotive direction.

Sand Switch

Sand will be applied to the rail in front of the leading axle of each truck when locomotive speed is less than 7 mph. Above 7 mph, manual sanding is not available. Lead axle sand will still function normally.

Bell Valve

Pull the valve handle to sound the bell. Push it in to shut off.

An option provides both the bell and horn to be sounded when the valve is operated.

Horn Valve

Pull the valve handle toward the operator to sound the horn.

Radio Location

This location is provided for the radio. Operation of the radio should be in accordance with railroad operating rules and procedures.

Air Gages

See AIR BRAKE EQUIPMENT section of this manual.

Brake Pipe Flow Indicator

See AIR BRAKE EQUIPMENT section of this manual.

Automatic Brake Valve Handle

See AIR BRAKE EQUIPMENT section of this manual.

Independent Brake Valve Handle

See AIR BRAKE EQUIPMENT section of this manual.

Brake Pipe Cut-Out Pilot Valve

See AIR BRAKE EQUIPMENT section of this manual.

MU2A or Dual Ported Cut-Out Cock

See AIR BRAKE EQUIPMENT section of this manual.

Brake Pipe Regulating Valve

See AIR BRAKE EQUIPMENT section of this manual.

Safety Control Foot Pedal

See Safety Control Foot Pedal section of this manual.

DEVICES ON ENGINE CONTROL PANEL (Fig. 3)

The Engine Control (EC) panel is located on the rear wall of the operator's cab, Fig. 3. Mounted on this panel are various switches, circuit breakers and operating devices used during locomotive operation.

OPERATING CONTROLS

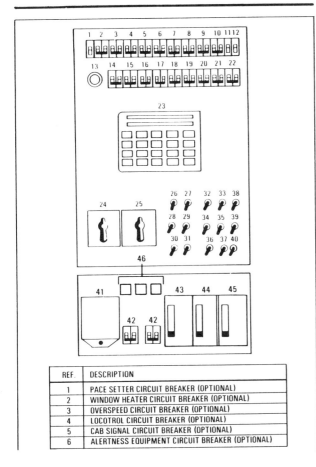

REF.	DESCRIPTION
1	PACE SETTER CIRCUIT BREAKER (OPTIONAL)
2	WINDOW HEATER CIRCUIT BREAKER (OPTIONAL)
3	OVERSPEED CIRCUIT BREAKER (OPTIONAL)
4	LOCOTROL CIRCUIT BREAKER (OPTIONAL)
5	CAB SIGNAL CIRCUIT BREAKER (OPTIONAL)
6	ALERTNESS EQUIPMENT CIRCUIT BREAKER (OPTIONAL)

REF	DESCRIPTION (CONT'D.)
7	WARNING LIGHT CIRCUIT BREAKER (OPTIONAL)
8	OSCILLATING LIGHT CIRCUIT BREAKER (OPTIONAL)
9	RADIO CIRCUIT BREAKER (OPTIONAL)
10	WATER COOLER CIRCUIT BREAKER (OPTIONAL)
11	FRONT HEADLIGHT CIRCUIT BREAKER
12	REAR HEADLIGHT CIRCUIT BREAKER
13	ENGINE STOP BUTTON
14	AUTOMATIC WATER DRAIN CIRCUIT BREAKER (OPTIONAL)
15	TOILET TANK HEATER CIRCUIT BREAKER (OPTIONAL)
16	FLANGE LUBRICATOR CIRCUIT BREAKER (OPTIONAL)
17	AIR DRYER CIRCUIT BREAKER (OPTIONAL)
18	RUNNING LIGHTS CIRCUIT BREAKER (ALL LIGHTS EXCEPT HEADLIGHTS)
19	AUTO FUEL CIRCUIT BREAKER (OPTIONAL)
20	FUEL PUMP CIRCUIT BREAKER
21	LOCAL CONTROL CIRCUIT BREAKER
22	BATTERY CHARGE AND COMPUTER CIRCUIT BREAKER
23	DIAGNOSTIC DISPLAY PANEL (DID)
24	ENGINE CONTROL SWITCH
25	MU HEADLIGHT SET-UP SWITCH
26	CROSSWALK LIGHT SWITCH
27	CONTROL COMPARTMENT LIGHT SWITCH
28	FRONT NUMBER LIGHT SWITCH
29	REAR NUMBER LIGHT SWITCH (OPTIONAL)
30	FRONT CLASS LIGHT SWITCH (OPTIONAL)
31	REAR CLASS LIGHT SWITCH (OPTIONAL)
32	NUMBER 1 MOTOR CUT-OUT
33	NUMBER 2 MOTOR CUT-OUT
34	NUMBER 3 MOTOR CUT-OUT
35	NUMBER 4 MOTOR CUT-OUT
36	NUMBER 5 MOTOR CUT-OUT (6-AXLE LOCOMOTIVE ONLY)
37	NUMBER 6 MOTOR CUT-OUT (6-AXLE LOCOMOTIVE ONLY)
38	SPEED SENSOR CUT-OUT SWITCH
39	LOCKED AXLE CUT-OUT SWITCH
40	DYNAMIC BRAKE CUT-OUT SWITCH (OPTIONAL)
41	BATTERY CHARGE RECEPTACLE (OPTIONAL)
42	WALL HEATER CIRCUIT BREAKER (ENGINEER'S AND HELPER'S)
43	CAB HEATER CIRCUIT BREAKER ENGINEER'S POSITION
44	CAB HEATER CIRCUIT BREAKER HELPER'S POSITION
45	AIR CONDITIONER CIRCUIT BREAKER (OPTIONAL)
46	AUTO FUEL INDICATING LIGHTS (OPTIONAL)

FIG. 3. ENGINE CONTROL PANEL.

Top Row of Circuit Breakers

The top row of circuit breakers on the EC panel are used for optional equipment or equipment that can be turned OFF when the unit is operating as a Trail unit. From left to right the circuit breakers and their functions are listed. Absence of one of these circuit breakers indicates that the locomotive is not equipped with that option.

Pace Setter (optional)
Window Heater (optional)
Overspeed (optional)
Locotrol (optional)
Cab Signal (optional)
Alertness Equipment (optional)
Warning Light (optional)
Oscillating Light (optional)
Radio (breaker is standard, equipment is optional)
Water Cooler (breaker is standard, equipment is optional)
Front Headlight
Rear Headlight.

Second Row of Circuit Breakers

The circuit breakers in the second row of circuit breakers are used for both standard and optional equipment, all of which MUST BE LEFT ON whenever the unit is operating as a Lead or Trail unit. From left to right the circuit breakers and their functions are listed:

Automatic Water Drain (optional)
Toilet Tank Heater (optional)
Flange Lubricator (optional)
Air Dryer (optional)
Running Lights (all lights except headlights)
Auto Fuel (optional)
Fuel Pump
Local Control
Battery Charge and Computer.

Additional equipment on the Engine Control panel is discussed below:

Engine Stop Button.

To shut down the engine, press the Engine Stop button.

Diagnostic Display Panel (DID)

See DIAGNOSTIC DISPLAY PANEL section of this manual.

Engine Control Switch

The Engine Control (EC) switch has four positions:

1. START - The Engine Start switch, see Engine Start Station, is effective only when the EC switch is in START. When the engine is running and the EC switch is in START position, engine speed is held at IDLE and power cannot be applied to the locomotive. The power plant is said to be "off the line." The alarm bell will not ring if the engine shuts down.

2. ISOLATE - When the engine is running and the EC switch is in the ISOLATE position, the engine speed is held at IDLE and power cannot be applied to the locomotive. The message "ISOLATED" will appear on the Diagnostic Display Panel. The alarm bell will sound if a fault occurs that will shut down the engine.

-223-

OPERATING CONTROLS

3. RUN - When the engine is idling and the locomotive is to be operated, the Engine Control (EC) switch must be moved to the RUN position.

NOTE: *If the EC switch is left in the RUN position when the diesel engine is shut down, the alarm bell will sound and a message will appear on the Diagnostic Display Panel.*

4. JOG - When the engine is shutdown and the locomotive is to be moved using battery power, the EC switch is moved to the JOG position.

NOTE: *JOG is optional equipment.*

MU Headlight Set-Up Switch

The MU Headlight Set-Up switch has five positions. Positioning of this switch is determined by location of the locomotive unit in the consist and whether the front of the locomotive unit is leading or trailing. Switch positions are as follows:

1. SINGLE OR MIDDLE UNIT - Place switch in this position on any locomotive unit operated singly or on all units, except the Leading or Trailing unit, when the locomotive consist is made up of more than one unit.

2. SHORT HOOD LEAD - LEADING UNIT - Place switch in this position when the Leading unit is operated with the short hood forward.

3. LONG HOOD LEAD - LEADING UNIT - Place switch in this position when the Leading unit is operated with the long hood forward.

4. SHORT HOOD TRAIL - TRAILING UNIT - Place switch in this position when the final Trailing unit is connected so its short hood trails.

5. LONG HOOD TRAIL - TRAILING UNIT - Place switch in this position when the final Trailing locomotive is connected so its long hood trails.

Crosswalk Light Switch

Control Compartment Light Switch

This switch turns on lights in Control Areas 1, 2, 3, 4 and 7.

Front Number Light Switch

Operates front number lights.

Rear Number Light Switch (Optional)

Operates rear number lights.

Front Class Light Switch (Optional)

Operates front class lights.

Rear Class Light Switch (Optional)

Operates rear class lights.

Traction Motor Cut-Out Switches - Pull to Throw

Number 1 Motor Cut-Out
Number 2 Motor Cut-Out
Number 3 Motor Cut-Out
Number 4 Motor Cut-Out
Number 5 Motor Cut-Out (six-axle locomotive only)
Number 6 Motor Cut-Out (six-axle locomotive only).

The Motor Cut-Out switches can be used to cut-out one or more traction motors. At the same time, power output of the locomotive may be reduced. See ALARMS, SAFEGUARDS, POWER DERATIONS AND SHUTDOWNS section of this publication.

CAUTION: *It is recommended that these switches be operated only with the Engine Control switch in START or ISOLATE position so the unit is isolated and the Throttle handle in IDLE.*

Under emergency conditions, the locomotive may be operated for a short period of time with one or more motors cut-out. Refer to railroad rules for specific details of operation.

Speed Sensor Cut-Out Switch

Cuts out the Speed Sensor signal on all traction motors that are cut-out.

NOTE: *A minimum of two motor speed sensors must be operating for the unit to load.*

NOTE: *When the Motor Speed Cut-Out switch is in the CUT-OUT position, the speed signals from all speed sensors on motors cut-out are ignored.*

Locked Axle Cut-Out Switch

Cuts out the Locked Axle Alarm.

Dynamic Brake Cut-Out Switch (optional)

Battery Charge Receptacle (optional)

Engineer's Wall Heater Circuit Breaker

Helper's Wall Heater Circuit Breaker

Cab Heater Circuit Breaker Engineer's Position

Cab Heater Circuit Breaker Helper's Position

Air Conditioner Circuit Breaker (optional)

OTHER OPERATOR CAB CONTROLS

Battery Switch (Fig. 4)

Located behind door below the Engine Control panel.

Emergency Brake Valve (Fig. 5)

Handle located at the short hood end of the operator cab, between the front cab access door and the nose cab access door. Pulling this handle causes an Emergency brake application and dropping of power.

OPERATING CONTROLS

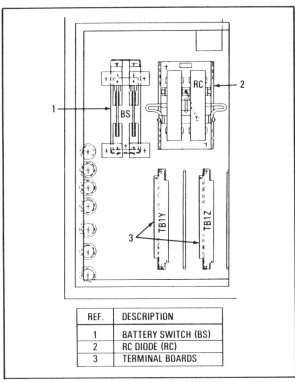

FIG. 4. BATTERY SWITCH COMPARTMENT BENEATH ENGINE CONTROL PANEL.

REF.	DESCRIPTION
1	BATTERY SWITCH (BS)
2	RC DIODE (RC)
3	TERMINAL BOARDS

Cab Heater/Defroster Controls - Engineer's and Helper's Positions (Fig. 5)

Cab heat and windshield defrost is regulated by a rotary switch on each heater that is labeled OFF-LO-MED-HI. Position this switch to the desired heat level from that heater. Each heater also has a toggle switch labeled "HIGH SPEED BLOWER SWITCH" which controls the speed of the heater blower. Position these switches to suit heating and defrosting needs as follows:

Toggle	Rotary	Result
OFF	OFF	Unit OFF
ON	OFF	Blower only at HIGH
OFF	LO	Blower at LOW Speed LOW Heat ON
ON	LO	Blower at HIGH Speed LOW Heat ON
ON	MED	Blower at HIGH Speed MED Heat ON
ON	HI	Blower at HIGH Speed HI Heat ON

The Over Heat circuit breaker located on the front of each heater and the Cab Heater circuit breakers located on the Engine Control panel must be ON for heaters to operate.

CAUTION: *To avoid overheating and tripping Heater breakers when Cab Heating System is in use, be sure return air inlet or heat outlets are not restricted.*

Windshield Wiper Valves (Fig. 6)

Located above the operator's and helper's positions.

Engineer's and Helper's Dome Lights (Fig. 6)

Located and controlled above the operator's and helper's positions.

Toilet and Nose Step Light Switch

Located on the back of the operator's console, this switch turns on the light in the nose cab.

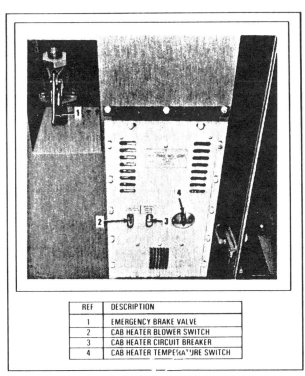

REF	DESCRIPTION
1	EMERGENCY BRAKE VALVE
2	CAB HEATER BLOWER SWITCH
3	CAB HEATER CIRCUIT BREAKER
4	CAB HEATER TEMPERATURE SWITCH

FIG. 5. EMERGENCY BRAKE VALVE AND HELPER'S HEATER (SHORT HOOD LEAD ARRANGEMENT).

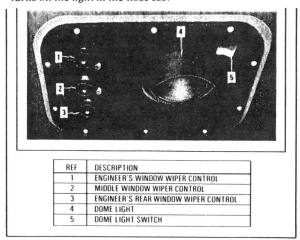

REF	DESCRIPTION
1	ENGINEER'S WINDOW WIPER CONTROL
2	MIDDLE WINDOW WIPER CONTROL
3	ENGINEER'S REAR WINDOW WIPER CONTROL
4	DOME LIGHT
5	DOME LIGHT SWITCH

FIG. 6. WINDOW WIPER CONTROLS AND DOME LIGHT.

Cab Air Conditioner (Optional)

An ON/OFF toggle switch located on the air conditioner turns the unit on. The Blower Speed switch can be used to run the fan at LOW or HIGH speed. The circuit breakers on the air conditioner and on the Engine Control panel, Fig. 3, must be ON and the battery charger running for the unit to operate.

DIAGNOSTIC DISPLAY PANEL (DID)

GENERAL INFORMATION

The DID panel is a fast and accurate means of communications between the locomotive operator and computers. The DID panel can be utilized in several ways:

1. If an abnormal operating condition (called a "FAULT") is detected, the computers will initiate the ALARM mode. In the ALARM mode, the computer uses the DID panel to alert the operator to the FAULT by displaying a description of the FAULT and, in some cases, ringing the Alarm bell. All FAULT messages are preceded by a four digit Fault Number.

2. The FAULT detected may require that certain operating restrictions be imposed on the locomotive as a means of protecting the locomotive's equipment. The locomotive computers impose the necessary restrictions and inform the operator of those restrictions through the DID panel in the form of SUMMARY messages.

3. A SUMMARY message on the display, informs the operator of the general status of the locomotive's operating condition, its computers, restriction placed on the locomotive due to faults and, in some cases, the status of the display itself. A SUMMARY message is not preceded by a four digit Code Number.

4. The FAULT is recorded in a FAULT "Log" for later review by maintenance personnel.

5. The operator can use the DID panel to review all active FAULTS and their related restrictions (SUMMARY messages). The DID panel also enables the operator to reset FAULTS, and attempt to return the locomotive to normal operation.

NOTE: *In some cases, the ability to reset certain FAULTS has been restricted to maintenance personnel in accordance with railroad selected options.*

THE DISPLAY

Message Windows

The Diagnostic Display (DID) panel has a two-line display window as described in Fig. 7.

Keys

Below the two-line display is a keypad with three rows of keys. Fig. 7 describes the use of these keys in Level 1. Other keys on the key pad are used on specific occasions, primarily for Level 2 maintenance operations.

NOTE: *Several levels of information access are available through the DID panel. Only Level 1 Operation of the panel is discussed in this publication.*

USING THE DISPLAY

Operation of the locomotive will not be interrupted or degraded and locomotive equipment will not be damaged if a mistake is made while using the DID panel in Level 1. Use of this panel by all responsible persons is encouraged.

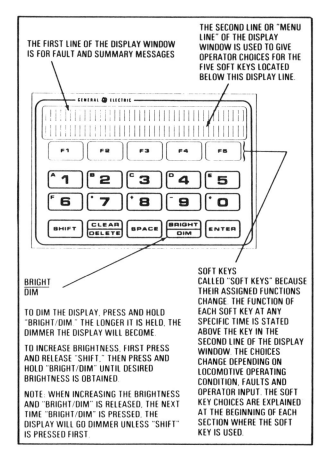

FIG. 7. DIAGNOSTIC DISPLAY PANEL.

MESSAGES AT LOCOMOTIVE START-UP

Certain SUMMARY messages are intended to inform the operator of the condition on the DID panel and the locomotive computers as they are powered-up. Several examples follow:

NOTE: *These are special SUMMARY messages which are not a result of FAULTS. They require no reset and are not stored in the FAULT log.*

This Display indicates that the power was applied to the system and the DID panel is functioning. **NOTE:** *This display will appear for 10 to 15 seconds while computers are starting.*	`Display Is Ready`
WAIT indicates that the CAB controller is starting to bring the control system "on-line" after power-up. **NOTE:** *Display of the WAIT message longer than 30 seconds indicates that the CAB controller is not able to bring the control system "on-line" and a problem may exist.*	`WAIT`
The READY display indicates all systems are running and the locomotive is READY to function normally.	`READY`
This display indicates that the Engine Control (EC) switch is in the ISOLATED position.	`ISOLATED`

OPERATING MODES IN LEVEL 1

After the locomotive computers have been powered-up and are operating normally, three modes of operation are available in Level 1:

1. READY mode
2. ALARM mode
3. FAULT mode.

READY Mode

READY indicates that all of the locomotive systems are functioning properly, and the locomotive is **"ready" to operate at full power.** READY can be displayed in one of three ways:

1. READY, appearing alone indicates that there have been no FAULTS detected, or reset.

2. "READY-Work Report Stored" indicates a FAULT has occurred, it has been reset, and all operating restrictions imposed by the FAULT have been removed.

3. Some FAULTS do not impose operating restrictions on the locomotive. When this type of FAULT occurs, "READY - Fault Message Stored" will be displayed.

NOTE: *As can be seen on the SUMMARY message list, Pages 56 and 57, these READY messages are the three lowest priority messages.* <u>*They will not be displayed if higher priority SUMMARY messages (operating restrictions) exist.*</u>

ALARM Mode

The computers check locomotive operation on a continuing basis. If an abnormal condition (FAULT) is detected, the ALARM mode may be initiated by the locomotive computers.

NOTE: *If the computer initiates the ALARM mode, when the DID panel is operating in any other mode, it will interrupt that mode to display the ALARM. When the ALARM mode is completed, the display will return to its previous operating mode.*

When the ALARM mode is initiated, a description of the problem will be given on the first line of the display in the form of a FAULT MESSAGE, the word "Silence" will appear on the second line of the display and, in most cases, an alarm bell will sound.

NOTE: *When any unit in the locomotive consist initiates an ALARM, the alarm bell on <u>all</u> locomotives will ring. All Dash 8 locomotives in the consist are notified of the ALARM through the SUMMARY message, "Alarm from Other Unit." If the <u>initiating unit</u> is a Dash 8 locomotive, a message describing the FAULT and "Silence" will appear on the Display Panel of that unit as described above. Pressing "Silence" on the <u>initiating unit</u> will quiet the ALARM on <u>all</u> trainlined <u>units</u>. The bell can only be silenced from the initiating unit therefore, "Silence" does not appear on any other units in the consist. See "Silence" soft key.*

"Silence" Soft Key

"Silence" is the only soft key that appears in the ALARM mode. It does not appear in any other mode of operation.

DIAGNOSTIC DISPLAY PANEL (DID)

When "Silence" is pressed OR if 30 seconds pass, the ALARM mode is terminated, the bell will stop ringing, the word "Silence" will disappear. The first line of the display will change from the FAULT message to show the operating restriction which has the greatest effect on the locomotive's ability to operate normally (highest priority SUMMARY message).

NOTE: *A few ALARMS are considered so serious that the bell cannot be silenced. In the cases of ENGINE SHUTDOWN, for example, no "Silence" soft key appears. The EC switch on the SHUTDOWN unit must be turned to the START position to silence the bell.*

FAULT Mode

As mentioned before, as a result of abnormal conditions (FAULTS), it may be necessary to protect the locomotive's equipment, by placing certain operating restrictions on the locomotive.

The FAULT mode of operation allows the operator to return the locomotive to the READY condition **unless** conditions exist that prohibit READY operation.

The restrictions imposed are displayed in the form of SUMMARY messages. In resetting FAULTS it is important to know the following about SUMMARY messages:

1. If a FAULT is reset, the operating restrictions imposed by it are removed and the related SUMMARY messages are no longer displayed.

2. Several FAULTS may impose the same operating restrictions and will therefore, result in the same SUMMARY message.

NOTE: *A SUMMARY message will only be displayed once (by priority) regardless of the number of active FAULTS which generate the same message.*

3. A FAULT may result in more than one SUMMARY message.

4. Under normal operating conditions, the highest priority SUMMARY message will be displayed. Highest priority being those conditions which have the greatest effect on the locomotive's ability to operate normally.

5. A list of SUMMARY messages by priority appears on Pages 56 and 57.

FAULT Mode Soft Keys

The following soft keys can be used by the operator to view SUMMARY and FAULT messages and to begin and to complete the reset procedure.

Soft Key Label	Explanation
Exit	Takes the DID panel out of the current operating mode.
Reset?	This soft key asks the operator, "Do you want to Reset?" (a FAULT). It can only appear when there Active FAULTS. Resetting a FAULT which has imposed operating restrictions is the only way to return the locomotive to the READY condition.
Reset? (Cont'd.)	Resetting a FAULT requires two steps: Pressing "Reset?" **initiates** the reset procedure. When "Reset?" is pressed, the most recent FAULT will be displayed with the choice of resetting that FAULT or looking at other FAULTS which have not been reset ("Active" FAULTS). **NOTE:** *"Reset" (without the question mark) must be pressed to complete the reset procedure.*
Reset	Pressing this key **completes** the reset procedure. Pressing "Reset" tells the computer this FAULT has been corrected, to remove all operating restrictions imposed by it and, if there are no other Active FAULTS, to return the locomotive to normal operation. When **all** Active FAULTS have been reset, the message "READY - Work Report Stored" will be displayed. If other Active FAULTS remain, the highest priority SUMMARY message will be displayed.

CAUTION: *Equipment damage may result - If a FAULT reoccurs soon after being reset, the operator should NOT attempt to reset the FAULT more than three (3) times until the cause of the FAULT has been determined and corrected.*

Soft Key Label	Explanation
Reset (Cont'd.)	**NOTE:** *If a FAULT causes power to be removed, the unit may not load after the FAULT is reset until the call for power is removed and again requested. This is done by momentarily placing the Engine Control (EC) switch in the ISOLATED position.* **NOTE:** *If a FAULT is Active (not reset), it will not reoccur. If a FAULT is reset and the problem not corrected, the FAULT will reoccur and the ALARM mode will be re-initiated.*
Older and Newer	FAULT messages are displayed in order of **most recent** first. The "Older" and "Newer" soft keys allow the operator to view "Older" and "Newer" Active FAULT messages respectively.
ShoMore and GoBack	SUMMARY messages are displayed in order of **highest priority**. "ShoMore" and "GoBack" allow the operator to review ALL SUMMARY messages (operating restrictions). Each time "ShoMore" is pressed, the next lower priority SUMMARY message will be displayed. Pressing "GoBack" will display the next higher priority SUMMARY message.

Soft Key Label	Explanation
ShoMore and GoBack (Cont'd.)	**NOTE:** *The choices "ShoMore" and "GoBack" are given only when there are lower or higher priority SUMMARY messages respectively.* **NOTE:** *If there is no key pad activity for 15 seconds, the display will change to show the highest priority SUMMARY message.*

EXAMPLE - LEVEL 1 OPERATION

NOTE: *The following example is intended to demonstrate DID operation, rather than show actual locomotive operating circumstances.*

Let us assume, for example, that READY is displayed (the locomotive is in the READY mode).

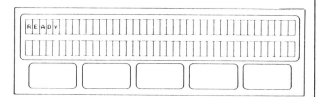

A hot diode condition in the main rectifiers (a FAULT) is detected and the ALARM mode is initiated.

The display will change to show the FAULT, the word "Silence" will appear, and in this case, the alarm bell will ring.

When "Silence" is pressed OR after 30 seconds pass, the ALARM mode is completed; the bell stops ringing, the word "Silence" disappears and the display changes to show the **highest priority** SUMMARY message.

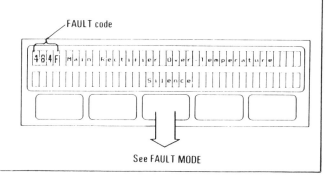

See FAULT MODE

Second Alarm

Next, let us assume that a bad outside air temperature sensor is detected. This FAULT occurred after the hot diode FAULT previously discussed and is therefore, a NEWER FAULT.

The ALARM mode is initiated and the display will change to show the bad temperature sensor FAULT.

NOTE: *This FAULT is not accompanied by a bell but "Silence" will appear.*

The procedure as previously described will be followed, the ALARM mode will be completed and the highest priority SUMMARY message will be displayed.

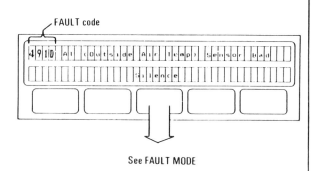

See FAULT MODE

FAULT Mode

The highest priority SUMMARY message is now displayed. "Won't Load: Hot Diodes" is the highest priority operating restriction placed on the locomotive as a result of the hot diode FAULT and the "Air Temperature Sensor Bad" FAULTS.

The operator now has two choices.

1. Press "Reset?" which will initiate the reset procedure, or
2. Press "ShoMore" to view all operating restrictions placed on the locomotive.

NOTE: *Four SUMMARY messages result from the "184F Main Rectifier Over-Temperature" FAULT. They are (highest to lowest priority):*

"Won't Load: Hot Diodes"
"Won't Load: Fault Message Stored"
"Won't Self-Load: Fault Message Stored"
"No Dynamic Brake: Fault Message Stored."

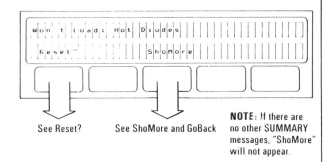

NOTE: If there are no other SUMMARY messages, "ShoMore" will not appear.

ShoMore and GoBack

"ShoMore" and "GoBack" allow the operator to review all restrictions placed on the locomotive as a result of Active FAULTS.

Each time "ShoMore" is pressed, the SUMMARY message **next lower** in priority to the message currently displayed is shown.

Each time "GoBack" is pressed, the SUMMARY message **next higher** in priority to the message currently displayed is shown.

NOTE: *If 15 seconds pass with no key pad activity, the display will change to show the highest priority SUMMARY message.*

NOTE: *If there are no lower priority SUMMARY messages, "ShoMore" will not appear. If there are no higher priority SUMMARY messages, "GoBack" will not appear.*

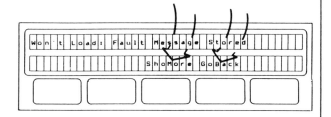

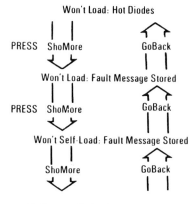

DIAGNOSTIC DISPLAY PANEL (DID)

Reset?

"Reset?" is the **first** step in the FAULT reset procedure.

When "Reset?" is pressed, the most recent (newest) FAULT message is displayed.

Older or Newer

"Newer" and "Older" allow the operator to look at all Active FAULTS and to select the FAULT to be reset.

Each time "Older" is pressed, the FAULT which occurred previous to the FAULT currently displayed will be shown.

Each time "Newer" is pressed, the FAULT which occurred after the FAULT currently displayed will be shown.

NOTE: *If there are no older FAULTS and "Older" is pressed, the message "(No Older Faults)" will appear. If there are no newer FAULTS and "Newer" is pressed, the message "(No Newer Faults)" will appear.*

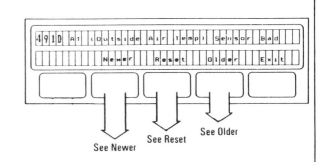

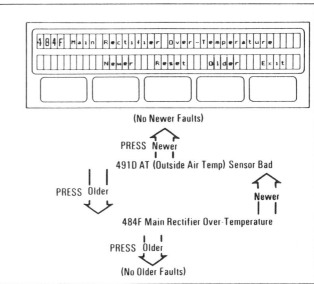

Reset

Several things happen when a FAULT is reset:

1. All operating restrictions imposed by the FAULT are removed.

2. If there are other Active FAULTS, the display will show the highest priority SUMMARY message of the remaining Active FAULTS.

3. If there are NO OTHER Active FAULTS, the display will change to show: "READY - Work Report Stored."

NOTE: *The SUMMARY message "READY - Work Report Stored" is for the locomotive maintainer. It tells the maintainer that problems have been encountered which should be investigated.*

DIAGNOSTIC DISPLAY PANEL (DID)

LIST OF SUMMARY MESSAGES

Highest Priority →

WAIT
WARNING! Air Compressor Does Not Pump
Won't Load: Locked Axle Detected
Automatic Water Drain Disabled
SHUTDOWN: Low Water Flow
SHUTDOWN: Low Oil Pressure
SHUTDOWN: Low Water Pressure
SHUTDOWN: Crankcase Overpressure
SHUTDOWN: Engine Overspeed
SHUTDOWN: Electrical Control Problem
Won't Crank: Electrical Control Problem
Engine Not Running
Can't Load Now: Too Much Cycling
Can't Charge Batteries Now: BRP Cycling
No Battery Charge: Elect. Control Prob.
No Battery Charge
Won't Battery Jog: Elect. Control Prob.
Can't Battery Jog: BKT in Wrong Position
Can't Self-Load: REV in Wrong Position
Won't Load: Overspeed Governor Problem
Won't Load: Aux. Alternator Field C/O
Won't Load: Side Door Open
Won't Load: Electrical Control Problem
Won't Load: Too Many Speed Sensors C/O
Won't Load: Waiting for Aux. Alternator
Won't Load: Hot Engine
Won't Load: Power Circuit Ground
Won't Load: Power Circuit Problem
Won't Load: Battery Charge Problem
Won't Load: Hot Diodes
Won't Load: MU Error

Won't Load: Fault Message Stored
Won't Crank: Fault Message Stored
Won't Battery Jog: Fault Message Stored
Won't Self-Load: Fault Message Stored
ISOLATED
Self-Load: LOAD CONDITIONS
Operating in STANDBY POWER Mode
No Dynamic Brake: Man. Tract. Motor C/O
No Dynamic Brake: Auto. Tract. Motor C/O
No Dynamic Brake: Elect. Control Prob.
No Dynamic Brake: Power Circuit Problem
No Dynamic Brake: Fault Message Stored
Warning: Locked Axle Alarm is Cut Out
Load Limited: PLS in Notch 7
Load Limited: PCS Trip
Load Limited: Low Oil Pressure
Load Limited: Low Water Pressure
Load Limited: Hot Engine
Load Limited: Cold Engine
Load Limited: Dirty Engine Air Filter
Load Limited: Traction Motors Cut Out
Load Limited: Trac. Motor Temp. Protection
Load Limited: Power Circuit Ground
Load Limited: Electrical Control Problem
May Reduce Load: Radiator Fan Cycling
May Reduce Load: Radiator Fan Problem
Wrong Wheel Dia./Overspeed Calibration
Alarm From Other Unit
Fault Log is Almost Full
READY - Fault Message Stored
READY - Work Report Stored
READY

Lowest Priority

©1987 GENERAL ELECTRIC COMPANY

-232-

The Schedule 26-L equipment, arranged for single-end, multiple-unit operation, is used on this locomotive. The principal parts are as follows:

AIR BRAKE EQUIPMENT ON THE CONTROL CONSOLE (Fig. 2)

26-C Brake Valve

This valve consists of two pieces: the automatic brake valve and the independent brake valve. The automatic valve regulates brake pipe pressure to control both locomotive and train brakes. The independent valve controls application and release of the locomotive brakes independent of the train brakes. The independent valve also controls the release of the automatic brake on the locomotive or locomotive consist without effecting the Automatic application on the rest of the train.

Automatic Brake Valve Handle (Fig. 8)

The Automatic Brake Valve handle has six positions:

1. RELEASE (RUNNING) position - This position charges the brake pipe and air brake equipment to release the automatic air brake on the locomotive and train after an Automatic application. This is accomplished by controlling air flow to the brake pipe as set by the regulating valve (on back of brake stand). The RELEASE position is at the extreme left of the quadrant and is the normal position when the automatic brake is not in use.

2. MINIMUM REDUCTION position - This position is located to the right of the RELEASE position where the Brake Valve handle reaches the first raised portion

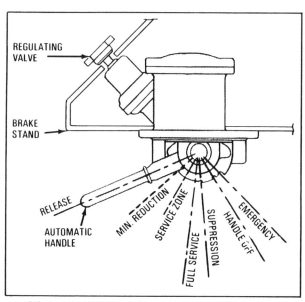

FIG. 8. AUTOMATIC BRAKE VALVE HANDLE POSITIONS.

of the quadrant. With the Brake Valve handle moved to this position, the Minimum Service application is obtained, which results in a four to six pound brake pipe reduction.

3. SERVICE positions - This sector of the Brake Valve handle movement is to the right of the MINIMUM REDUCTION position. Moving the handle from the left to right in this sector gradually increases the degree of brake application. At the extreme right of the sector, a Full Service brake application is obtained.

AIR BRAKE EQUIPMENT

4. SUPPRESSION position - This position is located with the handle against the second raised position of the quadrant, to the right of the RELEASE position. This position provides a Full Service brake application and, in addition, on locomotives equipped with overspeed control and safety control penalty brakes, these applications will be suppressed.

5. HANDLE-OFF position - This position is located by the quadrant notch to the right of the SUPPRESSION position. The handle is removable in this position. It must be placed in this position and removed on trailing units of a multiple-unit consist or on locomotives being towed "dead-in-train."

6. EMERGENCY position - This position is located to the extreme right of the brake valve quadrant. It is used for making a brake valve Emergency brake application.

When an Emergency application has occurred, the Automatic Brake Valve handle must be moved to the EMERGENCY position and left in this position until the equalizing reservoir gage hand indicates zero ("0") pressure and the Sand light is out. The Automatic Brake Valve handle then must be moved to the RELEASE position to recharge the brake pipe and release the brakes.

Independent Brake Valve Handle

The Independent Brake Valve handle applies and releases the brakes on the locomotive consist or releases, on the locomotive consist only, the Automatic brake application after an Automatic or Emergency application.

The independent brake valve has two positions: RELEASE and FULL APPLICATION, with the application zone between. The brake valve is of the self-lapping type which automatically maintains brake cylinder pressure when the application pressure reaches a value corresponding to the handle position. An Independent brake application can be released only by movement of the handle toward the RELEASE position. An Automatic Service or Emergency application can be released on the locomotive consist by depressing the Independent Brake Valve handle in the RELEASE position.

NOTE: *If independent brakes are applied, only minimum dynamic brake can be obtained.*

Brake Pipe Cut-Out Pilot Valve (Fig. 2)

This cock, also known as the "double-heading cock" is located on the front of the automatic brake valve. Push in the handle and turn to position for type of service. The IN position is used when the locomotive is operated as a Lead unit. The OUT position is used when the locomotive is operated as a Trail unit.

MU2A Valve or Dual Ported Cut-Out Cock (Fig. 2)

This is a two-position valve located on the side of the brake stand. It enables a locomotive equipped with 26-L brakes to be operated in multiple with locomotives having smaller type brake equipment.

The two-position MU2A valve has positions LEAD/DEAD and TRAIL and the Dual Ported cut-out cock has positions IN/OPEN and OUT/CLOSED.

1. LEAD/DEAD or IN/OPEN position is used when locomotive unit is operated singly or when it is the Lead unit of a multiple-unit consist. Position is also used when locomotive unit is hauled "dead-in-train."

2. TRAIL or OUT/CLOSED position is used to trail a Lead locomotive having 26-L brake equipment.

Duplex Air Gages (Fig. 2)

The following duplex (two hands) air gages are located on the operator's console.

Main Reservoir - Equalizing Reservoir - Red hand indicates Main Reservoir (MR) pressure; White hand indicates Equalizing Reservoir (ER) pressure.

Brake Cylinder - Brake Pipe - Red hand indicates locomotive Brake Cylinder (BC) pressure; White hand indicates Brake Pipe (BP) pressure.

Brake Pipe Air Flow Indicator (Optional) (Fig. 2)

Air flow in the Brake Pipe is indicated by the White hand. The Red hand is set by the operator as maximum brake pipe flow. When the flow is greater than that set, a Red light on the bottom of the indicator will appear.

AIR BRAKE EQUIPMENT IN AIR BRAKE COMPARTMENT

See Fig. 9 for location of equipment in the air brake compartment. Presence of equipment will depend on the options selected by a railroad. See the Air Piping Diagram for specific air brake valve locations.

AIR COMPRESSOR SAFETY VALVES

This valve is located in the piping to the first main reservoir at the long hood end of the fuel tank. It is set to open at 150 psi. An optional safety valve is located at the air outlet of the air compressor and is set to operate at 175 psi.

Air Dryer (Optional)

Operation of the air dryer requires:

1. The Air Dryer Circuit Breaker (ADCB), located on the engine control panel be ON, Fig. 3.

2. The air compressor be pumping (loaded).

3. Main reservoir pressure be at least 110 psi.

Inspect the humidity indicator located near the outlet of the air dryer. When the air dryer is operating properly, the humidity indicator will show a blue color. If the indicator is white, the air dryer is not operating properly and maintenance is required.

There should be a slight, continuous exhaust of air at the exhaust port of the purge valve on one dryer. After about one minute, there should be a short, loud discharge of air at the exhaust port of the opposite purge valve, followed by a slight, continuous exhaust of air. About one minute later, there should be a similar discharge of air at the opposite dryer. The change in the exhaust of air from one dryer to the other indicates the dryer is operating properly.

CUT-OUT COCKS

At specified inspection or maintenance periods, the following manually operated devices are used:

AIR BRAKE EQUIPMENT

1. Main Reservoir Cut-Out cock - Located on right side of locomotive near the main reservoir, Fig. 10.

2. Main Reservoir Drain cocks - One located on the end of each main reservoir, usually part of automatic drain valves, Fig. 11.

3. Air-Filter Drain cocks - Located on the main reservoir and auxiliary air filters, Fig. 10.

4. Control-Air Cut-Out cock - Located in air brake compartment, Fig. 9, Item 20.

5. Control-Air Reservoir Drain cock - Located in air brake compartment on rear wall, Fig. 9.

6. Brake Cylinder Cut-Out cocks - Located on right side beneath locomotive platform level (one for each truck), Fig. 12.

7. Air Compressor Governor Cut-Out cock - Located in air compressor compartment accessible from right side of locomotive, Fig. 13.

8. Bell, Horn and Window Wiper Cut-Out cock - Located in air brake compartment, Fig. 9.

9. Sander Control Cut-Out cocks - The front sander cut-out cocks are located in the air brake compartment, Fig. 9. The cut-out cocks for the rear sanders are located inside the radiator cab below the sand box on the left side of the locomotive.

10. Cut-Out Cocks and End Connections in each end of locomotive, Fig. 14:
    a. Brake Pipe Angle cocks or cut-out cock located behind end frame (BP)
    b. Main Reservoir Equalizing (MR)
    c. Actuating (ACT)
    d. Brake Cylinder Equalizing (Independent Application and Release) (AP).

11. Safety Control Cut-Out cock (optional) - Located in air brake compartment, Fig. 9, or in the nose cab (optional location). Cuts out safety control feature when closed. (See Air Piping Diagram for inclusion and specific location.)

12. Overspeed Control Cut-Out cock (optional) - Located in air brake compartment, Fig. 9. Cuts out overspeed control feature when closed.

13. Dead Engine cock - Located in air brake compartment as part of the air brake rack, Fig. 9.

ADJUSTING VALVES

Brake Pipe Regulating Valve (Fig. 2)

The brake pipe regulating valve, located on the control stand, automatically maintains a predetermined air pressure in the brake system. A clockwise movement of the adjusting handle increases the pressure setting. A counterclockwise movement decreases the pressure setting. Adjust to conform with railroad regulations.

Control Air Reducing Valve (Located in Control Area 7, Fig. 20)

This valve maintains a predetermined normal air pressure in the air pressure supply for operation of pneumatically operated control equipment. Clockwise adjustment of the adjusting screw increases pressure. Normal control air pressure is 80 lb.

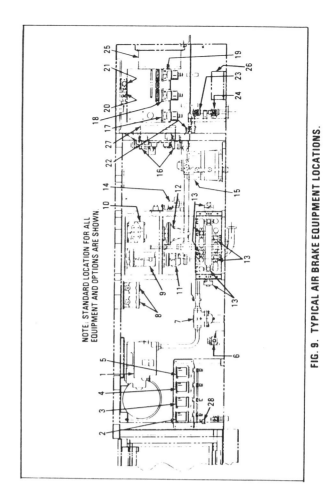

FIG. 9. TYPICAL AIR BRAKE EQUIPMENT LOCATIONS.

REF.	DESCRIPTION
1	26F CONTROL VALVE
2	PRESSURE SWITCH (DBCO OR BCPS)
3	PRESSURE SWITCH (IBS)
4	PRESSURE SWITCH (SPS)
5	PRESSURE SWITCH (PCS)
6	DEAD ENGINE FIXTURE
7	BRAKE PIPE CUT-OUT COCK AND FILTER
8	IN-LINE FILTERS (ACT. PIPE AND IND. APPL. AND RELEASE PIPE)
9	P2A BRAKE APPLICATION VALVE
10	A-1 CHARGING CUT-OFF PILOT VALVE
11	H-5 RELAY AIR VALVE
12	N-2 REGULATING VALVE
13	CHECK AND DOUBLE CHECK VALVES
14	DYNAMIC BRAKE MAGNET VALVE (DBM)
15	RELAY VALVE (J1.4, J1.14 OR J1.6.16)
16	SHORT HOOD END SANDER CONTROL VALVES (FSCV1 AND FSCV2)
17	MAGNET VALVE
18	MAGNET VALVE (OSV)
19	SAFETY CONTROL MAGNET VALVE (SCMV)
20	CONTROL AIR CUT-OUT COCK
21	CONTROL AIR CHECK VALVE NOTE: CONTROL AIR REGULATING VALVE AND GAGE ARE LOCATED IN CONTROL AREA 7, SEE FIG. 20.
22	PULSE AIR CUT-OUT COCK
23	FORWARD SAND CUT-OUT COCK
24	HORN, BELL AND WIPER CUT-OUT COCK
25	PULSE EQUIPMENT - TRAIN SENTRY (OPTIONAL)
26	PULSE SPEED RECORDER (OPTIONAL)
27	PULSE AIR BRAKE MODULE
28	CONTROL AIR RESERVOIR DRAIN VALVE

FIG. 9. TYPICAL AIR BRAKE EQUIPMENT LOCATIONS.

AIR BRAKE EQUIPMENT

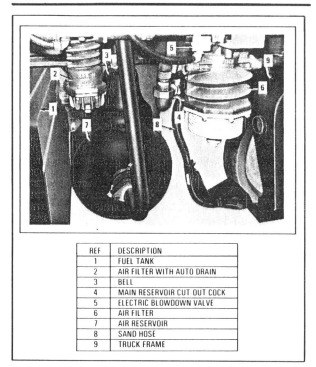

REF	DESCRIPTION
1	FUEL TANK
2	AIR FILTER WITH AUTO DRAIN
3	BELL
4	MAIN RESERVOIR CUT OUT COCK
5	ELECTRIC BLOWDOWN VALVE
6	AIR FILTER
7	AIR RESERVOIR
8	SAND HOSE
9	TRUCK FRAME

FIG. 10. TYPICAL MAIN RESERVOIR CUT-OUT COCK, MAIN AND AUXILIARY FILTERS AND DRAINS.

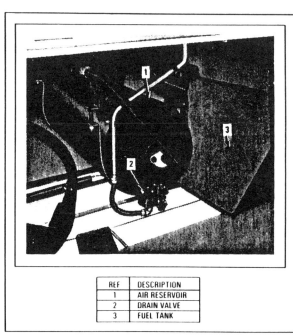

REF	DESCRIPTION
1	AIR RESERVOIR
2	DRAIN VALVE
3	FUEL TANK

FIG. 11. TYPICAL MAIN RESERVOIR DRAIN VALVE.

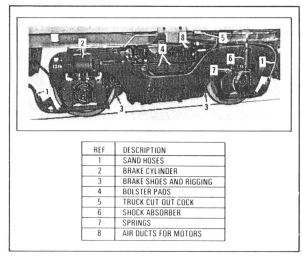

REF	DESCRIPTION
1	SAND HOSES
2	BRAKE CYLINDER
3	BRAKE SHOES AND RIGGING
4	BOLSTER PADS
5	TRUCK CUT OUT COCK
6	SHOCK ABSORBER
7	SPRINGS
8	AIR DUCTS FOR MOTORS

FIG. 12. TRUCK EQUIPMENT - FOUR-AXLE LOCOMOTIVE.

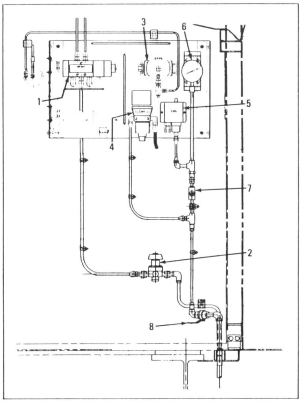

FIG. 13. AIR COMPRESSOR CONTROL PANEL, ENGINE AIR FILTER PRESSURE SWITCH AND ENGINE COOLING WATER FLOW CONTROL VALVE AND PRESSURE REGULATOR.

REF.	DESCRIPTION
1	ENGINE COOLING WATER FLOW CONTROL VALVE (WFMV)
2	PRESSURE REGULATOR
3	ENGINE AIR FILTER PRESSURE SWITCH (EFPS)
4	COMPRESSOR MAGNET VALVE (CMV)
5	COMPRESSOR GOVERNOR SWITCH (CGS)
6	MAIN RESERVOIR PRESSURE GAGE
7	GAGE AND SWITCH TEST FITTING
8	COMPRESSOR CUT-OUT COCK

FIG. 13. AIR COMPRESSOR CONTROL PANEL, ENGINE AIR FILTER PRESSURE SWITCH AND ENGINE COOLING WATER FLOW CONTROL VALVE AND PRESSURE REGULATOR.

AIR BRAKE EQUIPMENT

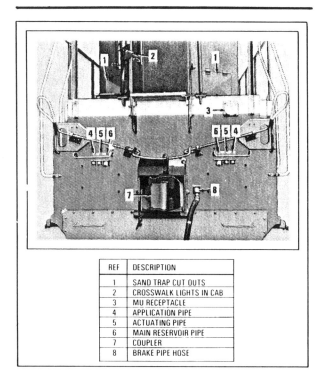

REF	DESCRIPTION
1	SAND TRAP CUT OUTS
2	CROSSWALK LIGHTS IN CAB
3	MU RECEPTACLE
4	APPLICATION PIPE
5	ACTUATING PIPE
6	MAIN RESERVOIR PIPE
7	COUPLER
8	BRAKE PIPE HOSE

FIG. 14. AIR BRAKE END CONNECTIONS.

OTHER EQUIPMENT

CONTROL COMPARTMENT EQUIPMENT (Fig. 15)

An equipment locker located at the long hood end of the operator cab, and the auxiliary cab house most of the control equipment. The control locker in the operator cab is called Control Area 1. The auxiliary cab, located directly behind the operator cab, houses Control Areas 2, 3 and 4.

In addition, control equipment is located in Control Areas 6 and 7 which are located on the left (B) side and Control Area 8 located on the right (A) side of the locomotive. These Control Areas are accessible from track level. Control Area 9 is located in radiator cab.

Following is a description of each of the control areas:

Engine Control Panel

Equipment on the EC panel is described earlier in this manual, see pages 15 through 23.

Control Area 1

Control locker at the long hood end of the operator cab, Fig. 16.

Control Equipment Areas Located in the Auxiliary Cab (Control Areas 2, 3 and 4)

> **WARNING:** *High voltage is present in this compartment when locomotive is under load. When the door to this compartment is opened, the Door Interlock Switch (DIS) will trip causing the unit to drop power. As a safety precaution, open the Auxiliary Alternator Cut-Out switch (BFCO) located inside Control Area 1, Fig. 16, before entering this compartment.*

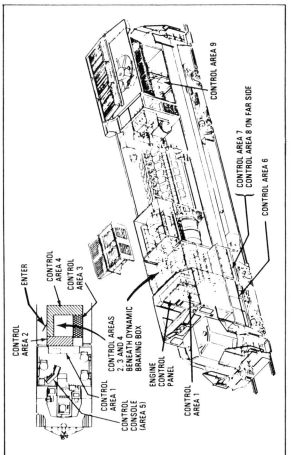

FIG. 15. LOCATION OF CONTROL AREAS.

OTHER EQUIPMENT

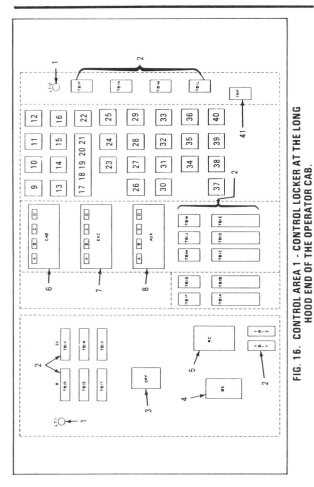

FIG. 16. CONTROL AREA 1 - CONTROL LOCKER AT THE LONG HOOD END OF THE OPERATOR CAB.

REF.	DESCRIPTION
1	COMPARTMENT LIGHTS
2	TERMINAL BOARDS
3	COMPUTER POWER FILTER (CPF)
4	BATTERY SWITCH (BS)
5	RC DIODE (RC)
6	CAB CONTROLLER
7	EXC CONTROLLER
8	AUX CONTROLLER
9	A, B, C SPEED VALVE RELAY (ABCR)
10	ALARM BELL RELAY (BLR)
11	DYNAMIC BRAKE RELAY (BR1)
12	COMPRESSOR LINE RELAY (CRL)
13	PNEUMATIC POWER CONTROL RELAY (PCR)
14	FUEL PUMP RELAY (FPR)
15	DV SHUT-DOWN RELAY (DVR)
16	RADIATOR FAN BYPASS RELAY (RFBR)
17	SELF-LOAD BOX TOGGLE SWITCH (LBTS)
18	LOAD BOX SELECTOR SWITCH (LBSS)
19	DIAGNOSTIC ACCESS SWITCH (DAS)
20	AUX. ALTERNATOR FIELD CUT-OUT SWITCH (BFCO)
21	RADIATOR FAN REVERSE SWITCH (FRB)
REF.	DESCRIPTION
22	BAROMETRIC PRESSURE TRANSDUCER (BPT)
23	WHEELSLIP ALARM RELAY (WSR)
24	LOCKED AXLE ALARM RELAY (LAR)
25	WATER FILL RELAY (WFR) (OPT.)
26	FUEL SAVER ELAPSED TIME CLOCK (CLOCK) (OPT)
27	FUEL SAVER RELAY (FSR) (OPT)
28	BATTERY JOG RELAY (JGR) (CPT)
29	WATER DRAIN ENABLE RELAY (DER) (OPT.)
30	FSR AUXILIARY RELAY (FSRA) (OPTIONAL)
31	PACE SETTER RELAY (PSR) (OPT.)
32	CAM W RELAY (CWR) (OPT.)
33	PACE SETTER AUXILIARY RELAY (GFA) (OPT.)
34	OSCILLATING HEADLIGHT RELAY (OHR) (OPT.)
35	OSC. HDLT. AUX. RELAY (AHRA) (OPT.)
36	PNEUMATIC PWR. CONTROL AUX. RELAY (PCRA) (OPT)
37	CAB SIG. FWD. DIR. RELAY (FDR) (OPT.)
38	CAB SIG. REV. DIR. RELAY (RDR) (OPT.)
39	OR AUTO FUEL NO FLOW RELAY (AFNR) (OPT.)
40	AUTO FUEL HIGH LEVEL RELAY (AFHR) (OPT.)
41	AUTO FUEL PUMP RELAY (AFPR) (OPT.)
42	TRAINLINE RESISTOR PANEL (TRP)

Control Area 2

Equipment located on the short hood end wall, Fig. 17.

Control Area 3

Equipment located on the left (B) side wall, Fig. 18.

Control Area 4

Equipment located on the long hood end wall and the right (A) side wall, Fig. 19.

Control Area 5 - Operator Control Console

Equipment located on the control console, is described earlier in this manual, see pages 4 through 15.

Control Area 6

Located on the left (B) side of the locomotive, beneath the short hood and operator cab. It is reserved for Locotrol, Cab Signal Equipment or other optional equipment.

Control Area 7

Located on the left (B) side of the locomotive, Fig. 20.

Control Area 8

Located on the right (A) side of the locomotive adjacent to the air brake compartment, Fig. 21.

Control Area 9

Located in the radiator cab on the left (B) side of the locomotive, Fig. 22.

OTHER EQUIPMENT

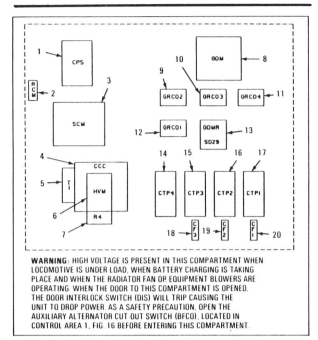

FIG. 17. CONTROL AREA 2 - EQUIPMENT LOCATED ON SHORT HOOD END WALL.

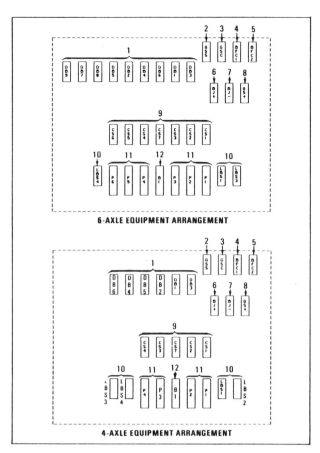

FIG. 18. CONTROL AREA 3 - EQUIPMENT LOCATED ON LEFT-SIDE WALL.

REF.	DESCRIPTION
1	COMPUTER POWER SUPPLY (CPS)
2	RECORDER CURRENT MODULE (RCM) (OPT.)
3	SIGNAL CONDITIONING UNIT (SCM)
4	CRANK COMMUTATING CAPACITOR (CCC)
5	VOLTAGE/HZ FEEDBACK TRANSFORMER (T1)
6	HIGH VOLTAGE INTERFACE UNIT (HVM)
7	CCC CAPACITOR BLEED RESISTOR (R4)
8	GROUND DETECTION UNITS (GDM)
9	GROUND RELAY CUT-OUT SWITCH #2 (GRCO2)
10	GROUND RELAY CUT-OUT SWITCH #3 (GRCO3)
11	GROUND RELAY CUT-OUT SWITCH #4 (GRCO4)
12	GROUND RELAY CUT-OUT SWITCH #1 (GRCO1)
13	CRANKING THYRISTOR PANEL #4 (CTP4)
14	CRANKING THYRISTOR PANEL #3 (CTP3)
15	CRANKING THYRISTOR PANEL #2 (CTP2)
16	CRANKING THYRISTOR PANEL #1 (CTP1)
17	CRANKING FUSE #3 (CF3)
18	CRANKING FUSE #2 (CF2)
19	CRANKING FUSE #1 (CF1)

FIG. 17. CONTROL AREA 2 - EQUIPMENT LOCATED ON SHORT HOOD END WALL.

REF.	DESCRIPTION
1	EXTENDED RANGE BRAKING CONTACTORS (IF USED): 4-AXLE LOCOMOTIVES, DB1 THRU DB6 INCL. 6-AXLE LOCOMOTIVES, DB1 THRU DB9 INCL.
2	ENGINE CRANK SEQUENCE CONTACTOR (GSS)
3	ENGINE CRANK CONTACTOR (GSC)
4	TRACT. ALT. FIELD EXCITING CONTACTOR (BFC1)
5	TRACT. ALT. FIELD EXCITING CONTACTOR (BFC2)
6	BATTERY JOG PLUS CONTACTOR (BJ+)
7	BATTERY JOG NEGATIVE CONTACTOR (BJ−)
	NOTE: BATTERY JOG IS AN OPTION.
8	ENGINE CRANK CONTACTOR (GS+)
9	CURRENT SHUNTS/SIGNAL CONDITIONING UNITS: 4-AXLE LOCOMOTIVES, CS1, 2, 3, 4 AND 7. 6-AXLE LOCOMOTIVES, CS1 THRU CS7 INCL.
10	SELF-LOAD BOX CONTACTORS: 4-AXLE LOCOMOTIVES, LBS1 THRU CS4 INCL. 6-AXLE LOCOMOTIVES, LBS1, 3 AND 4
11	POWER CONTACTORS: 4-AXLE LOCOMOTIVES, P1, 2, 3 AND 4 6-AXLE LOCOMOTIVES, P1 THRU P6 INCL.
12	DYNAMIC BRAKING CONTACTOR (B1)

WARNING: HIGH VOLTAGE IS PRESENT IN THIS COMPARTMENT WHEN LOCOMOTIVE IS UNDER LOAD, WHEN BATTERY CHARGING IS TAKING PLACE AND WHEN THE RADIATOR FAN OR EQUIPMENT BLOWERS ARE OPERATING. WHEN THE DOOR TO THIS COMPARTMENT IS OPENED, THE DOOR INTERLOCK SWITCH (DIS) WILL TRIP CAUSING THE UNIT TO DROP POWER. AS A SAFETY PRECAUTION, OPEN THE AUXILIARY ALTERNATOR CUT-OUT SWITCH (BFCO), LOCATED IN CONTROL AREA 1, FIG. 16 BEFORE ENTERING THIS COMPARTMENT.

FIG. 18. CONTROL AREA 3 - EQUIPMENT LOCATED ON LEFT-SIDE WALL.

OTHER EQUIPMENT

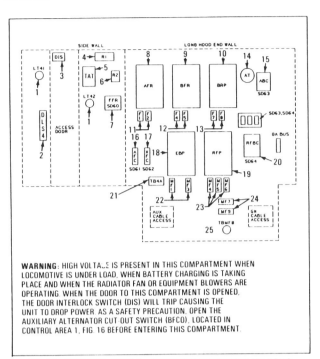

WARNING: HIGH VOLTAGE IS PRESENT IN THIS COMPARTMENT WHEN LOCOMOTIVE IS UNDER LOAD, WHEN BATTERY CHARGING IS TAKING PLACE AND WHEN THE RADIATOR FAN OR EQUIPMENT BLOWERS ARE OPERATING. WHEN THE DOOR TO THIS COMPARTMENT IS OPENED, THE DOOR INTERLOCK SWITCH (DIS) WILL TRIP CAUSING THE UNIT TO DROP POWER. AS A SAFETY PRECAUTION, OPEN THE AUXILIARY ALTERNATOR CUT-OUT SWITCH (BFCO), LOCATED IN CONTROL AREA 1, FIG. 16 BEFORE ENTERING THIS COMPARTMENT.

FIG. 19. CONTROL AREA 4 - EQUIPMENT LOCATED ON LONG HOOD END WALL AND THE RIGHT-SIDE WALL.

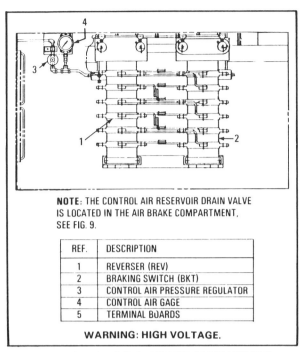

NOTE: THE CONTROL AIR RESERVOIR DRAIN VALVE IS LOCATED IN THE AIR BRAKE COMPARTMENT, SEE FIG. 9.

REF.	DESCRIPTION
1	REVERSER (REV)
2	BRAKING SWITCH (BKT)
3	CONTROL AIR PRESSURE REGULATOR
4	CONTROL AIR GAGE
5	TERMINAL BOARDS

WARNING: HIGH VOLTAGE.

FIG. 20. CONTROL AREA 7 - LOCATED ON THE LEFT SIDE OF THE LOCOMOTIVE.

REF.	DESCRIPTION
1	COMPARTMENT LIGHTS
2	DROP LIGHT SOCKET (DLS4)
3	DOOR INTERLOCK SWITCH (DIS)
4	ALTERNATOR FIELD CLIPPER RESISTOR (R1)
5	ALTERNATOR FIELD CLIPPER PANEL (TAT)
6	AUX ALT FIELD FLASHING RESISTOR (R2)
7	FIELD FLASHING RELAY (FFR)
8	TRACTION ALTERNATOR FIELD REGULATOR (AFR)
9	AUXILIARY ALTERNATOR FIELD REGULATOR (BFR)
10	BATTERY CHARGER REGULATOR (BRP)
11	FUSES (F1 AND F2)
12	FUSES (F4 AND F5)
13	FUSES (F7 AND F8)
14	AMBIENT AIR TEMPERATURE SENSOR (AT)
15	ALTERNATOR BLOWER CONTACTOR (ABC)
16	ALTERNATOR FIELD CONTACTOR (AFC)
17	AUXILIARY ALTERNATOR FIELD CONTACTOR (XFC)
18	EQUIPMENT BLOWER MOTOR DRIVE REGULATOR (EBP)
19	RADIATOR FAN MOTOR DRIVE REGULATOR (RFP)
20	RADIATOR FAN BYPASS CONTACTOR (RFBC)
21	POWER TERMINAL BOARD
22	EQUIPMENT BLOWER FUSES (MF1 AND MF3)
23	RADIATOR FAN FUSES (MF4, MF5 AND MF6)
24	COMPRESSOR DRIVE MOTOR FUSES (MF7 AND MF9)
25	MOTOR FUSE TERMINAL BOARD (TBMF8)

FIG. 19. CONTROL AREA 4 - EQUIPMENT LOCATED ON LONG HOOD END WALL AND THE RIGHT-SIDE WALL.

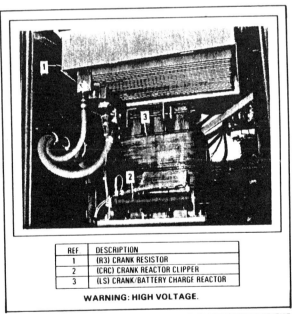

REF.	DESCRIPTION
1	(R3) CRANK RESISTOR
2	(CRC) CRANK REACTOR CLIPPER
3	(LS) CRANK/BATTERY CHARGE REACTOR

WARNING: HIGH VOLTAGE.

FIG. 21. CONTROL AREA 8 - LOCATED ON THE RIGHT SIDE OF THE LOCOMOTIVE ADJACENT TO THE AIR BRAKE COMPARTMENT.

OTHER EQUIPMENT

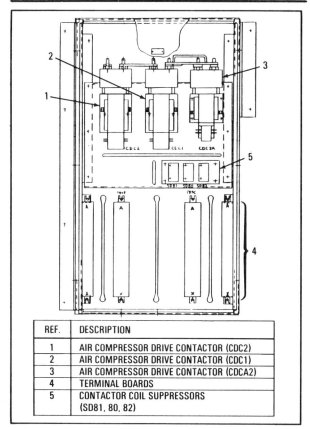

REF.	DESCRIPTION
1	AIR COMPRESSOR DRIVE CONTACTOR (CDC2)
2	AIR COMPRESSOR DRIVE CONTACTOR (CDC1)
3	AIR COMPRESSOR DRIVE CONTACTOR (CDCA2)
4	TERMINAL BOARDS
5	CONTACTOR COIL SUPPRESSORS (SD81, 80, 82)

FIG. 22. CONTROL AREA 9 - LOCATED IN THE RADIATOR CAB ON THE LEFT SIDE OF THE LOCOMOTIVE.

EQUIPMENT BLOWERS AND RADIATOR FAN (Fig. 23)

The Series-8 locomotive uses electric motor-driven traction motor blowers, one motor-driven alternator blower and a motor-driven radiator fan.

The speed of the traction motor blowers and the radiator fan are controlled by solid-state electronics, packaged in Replaceable Units, or RUs which are located in Control Area 4, Fig. 19. This type of control reduces auxiliary loads on the diesel engine since the blowers or fan run only when cooling is required, saving fuel.

Only the alternator blower is not speed-controlled by solid-state electronics. Its speed is directly proportional to engine speed.

A Fan Reverse switch, located in Control Area 1, Fig. 16, can be used to operate the radiator fan in reverse direction for a period of 60 seconds. This is to help clear leaves and debris which has accumulated on the inlet screens and radiators. This switch is intended for use by maintenance personnel.

NOTE: *If cooling water temperature is below 150 F, the blowers go to full speed.*

NOTE: *If ambient temperature is above 130 F, blowers go to full speed.*

NOTE: *If the radiator fans are not operated for a period of 30 minutes, the controllers will automatically operate them at full speed for a period of 10 seconds to prevent bearing brinelling.*

NOTE: *If fans have been cycling excessively, the fans will go to full speed.*

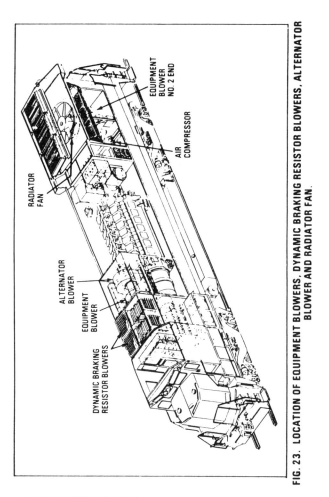

FIG. 23. LOCATION OF EQUIPMENT BLOWERS, DYNAMIC BRAKING RESISTOR BLOWERS, ALTERNATOR BLOWER AND RADIATOR FAN.

AIR COMPRESSOR

The Series-8 locomotive, built in 1987, uses an air compressor driven by an electric motor. Motor speed and compressor loading are controlled by the EXC Controller. The Compressor Governor Switch (CGS), located on the compressor control panel, monitors main reservoir pressure and provides a pressure signal to EXC. EXC, in turn, energizes the compressor drive contactor to start the air compressor drive motor. After 2 seconds, EXC de-energizes the Compressor Magnet Valve (CMV) to load the compressor. Speed of the air compressor drive motor is also monitored. If EXC has commanded the drive motor to start, but motor speed is not within limits, a FAULT will be logged, and the SUMMARY message "WARNING! Air Compressor Does Not Pump" will be displayed.

Engine Start Station and START Switch (Fig. 24)

The Engine Start Station is located in the engine cab next to the main traction alternator. It consists of an engine PRIME/START switch, which is used to start the diesel engine, and an ENGINE STOP button.

Gage Panel

An optional gage panel located near the engine start station houses gages which monitor turbo air pressure, fuel-oil pressure, lube-oil pressure at the governor and lube-oil pressure at the oil pump discharge.

OTHER EQUIPMENT

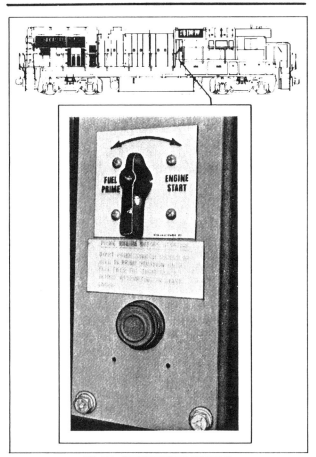

FIG. 24. ENGINE START STATION.

FIG. 25. ENGINE CONTROL GOVERNOR.

DIESEL ENGINE CONTROL GOVERNOR (Fig. 25)

The Diesel Engine Control Governor's primary function is to maintain speed of the diesel engine as called for by the Throttle handle notch setting in the lead locomotive controller. Engine rpm is maintained under a full range of loads. The governor also monitors the engine oil and water pressures, and will modulate the load and engine speed or, if necessary, shut the engine down if either pressure should fall below preset limits. Intake manifold air pressure is also monitored and the locomotive control system and the governor will limit the fuel available to the engine if the air pressure is below that required for complete combustion.

Engine Speed to Throttle Handle Position

Throttle Handle Position	Engine Notch	Engine RPM
Idle	(see Note 1)	
1	1	437-444
2	2	567-594
3	3	705-732
4	4	758-786
5	6	884-892
6	6	884-892
7	7	991-998
8 (see Note 2)	8	1047-1054

In Dynamic Braking, engine speed depends on the braking effort requested (position of Braking handle) and locomotive speed.

NOTE 1: *HIGH IDLE, REGULAR IDLE, LOW IDLE and LOW LOW IDLE:*

HIGH IDLE   567-594
Regular IDLE   437-444
LOW IDLE   324-352
LOW LOW IDLE 266-274

The locomotive control system will automatically reduce engine speed to LOW IDLE or LOW LOW IDLE based on the following requirements:

Reverse handle centered and locomotive NOT in Self-Load.

OR

Reverse handle in FWD or REV for more than five minutes with Throttle in IDLE and Braking handle in OFF.

AND

Battery charger current and voltage within certain limits based on the TIME current and voltages are within those limits.

AND

Engine cooling water and oil temperature within certain limits.

NOTE 2: *At certain locomotive speeds, with the Throttle handle in Notch 8, engine RPM will automatically be reduced from Notch 8 speed to Notch 7 speed while maintaining Notch 8 power. This occurs only on locomotives equipped with 16 cylinder engines (see GENERAL DATA section). This reduction occurs within a locomotive speed range which is determined by locomotive model and gearing.*

MISCELLANEOUS EQUIPMENT

1. Handbrake - Located on outside of nose compartment, Fig. 1, Item 4.

2. Emergency Fuel Cut-Off System, Figs. 3, 24 and 28. In an emergency, any one of four electric push-buttons may be depressed momentarily to cut off fuel delivery and shut down the engine. One of these buttons is located on each side of the locomotive platform near the fuel tank. The third and fourth buttons are located on the Engine Control (EC) panel and at the Start Station and are normally used for shutting down the engine.

NOTE: *The Emergency Cut-Off button is used to shut down the engine on the local units only. The SHUTDOWN position of the Throttle handle on the Master Controller will shut down the engines on all units of the consist simultaneously.*

3. Toilet (optional) - Located in the nose cab.

4. Water Cooler and Refrigerator (optional) - Located in the access to the nose cab.

GAGES AND MEASURING DEVICES
GEJ 6720

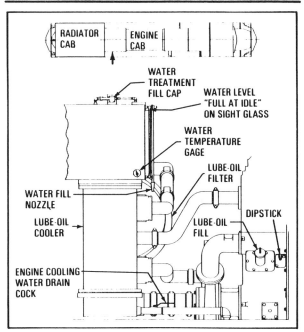

FIG. 26. ENGINE COOLING WATER SIGHT GLASS.

PRESSURE AND TEMPERATURE GAGES

1. Control Air Gage - Located in Control Area 7, Fig. 20. Normal control air pressure is 80 psi.

NOTE: *The following values are nominal due to the effect of varying conditions.*

2. Water Temperature Gage - Located on the right side of the water storage tank, Fig. 26. Normal operating temperature is 188-200 F.

FIG. 27. DIESEL ENGINE LUBE-OIL DIPSTICK AND FILL.

OTHER GAGES

1. Engine Lubricating-Oil Dipstick - Located on both sides of the engine near the lube-oil fill, Fig. 27. The stick is marked FULL and LOW. Proper level with the engine idling is between FULL and LOW.

NOTE: *Overfilling will cause engine to shutdown from excessive crankcase pressure.*

2. Fuel-Oil Sight Glasses - Mounted on both sides of the main fuel tank, Fig. 28, to indicate the level of fuel in the tanks.

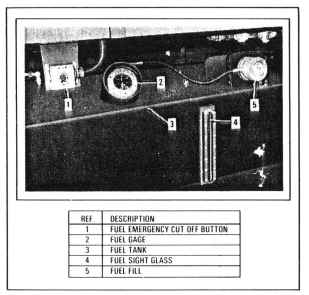

REF.	DESCRIPTION
1	FUEL EMERGENCY CUT OFF BUTTON
2	FUEL GAGE
3	FUEL TANK
4	FUEL SIGHT GLASS
5	FUEL FILL

FIG. 28. "A" SIDE OF FUEL TANK (TYPICAL).

3. Cooling Water - A water level sight glass mounted on the right side of the cooling water storage tank, Fig. 26, indicates the level of the cooling water. Markings near the sight glass indicate the proper level for various conditions of the system.

When filling the system or adding water treatment compound, proceed according to instructions mounted at the water storage tank area near the fill cap. Do not overfill.

WARNING: *To avoid personal harm from water burns, never remove the water fill cap when the water level is above FULL AT IDLE mark. If over-full, open manual drain valve to reduce the water to a safe level.*

4. Compressor Lube Oil, Fig. 29 - A dipstick, located near the fill cap, can be used to determine the oil level in the compressor crankcase.

5. Governor Oil-Level Sight Glass - Located on the left side of the engine near the traction generator, Fig. 25. Oil level must be visible at mark on the sight glass when the engine is running at idle.

CAUTION: *To prevent serious equipment damage, never start an engine until the governor has been properly serviced with lube oil.*

GAGES AND MEASURING DEVICES

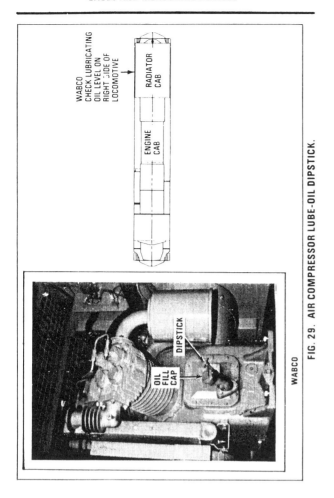

FIG. 29. AIR COMPRESSOR LUBE-OIL DIPSTICK.

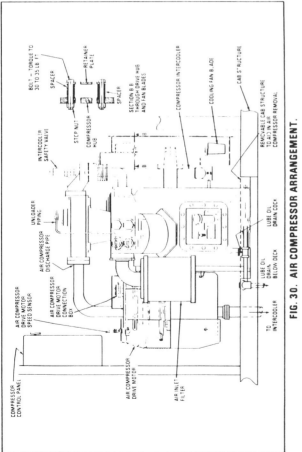

FIG. 30. AIR COMPRESSOR ARRANGEMENT.

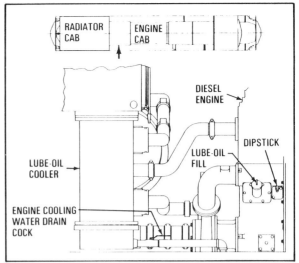

FIG. 31. ENGINE COOLING WATER DRAIN.

The cooling water system may be drained by opening the main water drain valve on the right side of the locomotive near the lube-oil pump, Fig. 31.

An optional Automatic Water Dump System will dump the engine cooling water when water temperature is below 40 F. A thermostat actuates, tripping the solenoid in the water drain valve. This opens the automatic drain valve and permits the rapid draining of the cooling water.

This system also has a Control switch located under the water tank. This switch can be used to fill the system with cold water, and to test the water dump valve.

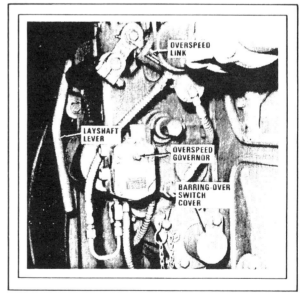

FIG. 32. ENGINE OVERSPEED SYSTEM AND BARRING-OVER SWITCH.

BARRING-OVER SWITCH (Fig. 32)

A Barring-Over switch is located under the cover of the diesel engine barring-over feature behind the engine overspeed governor. This switch prevents the engine from being cranked while engine barring-over procedure is in progress or if the cover has been left off.

-243-

ALARMS, SAFEGUARDS, POWER DERATIONS AND SHUTDOWNS

EMERGENCY SANDING

Emergency sanding is automatically applied in FORWARD and REVERSE directions during all Emergency brake applications for a sufficient time to stop the train. In multiple-unit operation, emergency sanding is applied to all units, regardless of whether they are equipped with pneumatic or electro-pneumatic sanding equipment.

NOTE: *Customer options may vary the operation of this switch.*

ENGINE AIR FILTER PRESSURE SWITCH (EFPS) (Fig. 13)

The Engine Air Filter Pressure Switch (EFPS) monitors air pressure drop across the engine air filters. When the Engine Air Filter switch operates, engine rpm follows Throttle handle and maximum power is limited to Notch 6.

GROUND CUT-OUT SWITCHES (Fig. 17)

Four Ground Cut-Out switches are mounted in Control Area 2 of the Series-8 locomotive.

These are two-pole switches which connect sensing circuits to detect ground leakage current in the following circuits:

1. Propulsion circuit (GRCO1)
2. Excitation supply circuit (GRCO2)
3. Auxiliary motor supply circuit (GRCO3)
4. Battery charging circuit (GRCO4).

One pole of each switch is used to remove the connection from the locomotive frame (chassis ground) to the ground detection circuitry. This is used to remove the "known" ground when performing insulation tests on the locomotive circuits, or to remove the "known" ground when troubleshooting for ground faults.

One pole of each switch is used to disable control circuits with the switch open.

THE LOCOMOTIVE WILL NOT LOAD WITH ANY OF THE GROUND CUT-OUT SWITCHES OPEN!

Only the propulsion circuit ground detector will derate locomotive performance based on ground leakage. Propulsion buss voltage is reduced proportional to ground leakage current as follows:

Ground Current Leakage (amperes)		Result
Motoring or Self-Load	Dynamic Brake	
0 to 1/2	0 to 1/4	Causes no deration.
1/2 to 1	1/4 to 1/2	Is the range which will derate propulsion buss voltage from no deration at 1/4 ampere to full deration at 1/2 ampere.
Above 1	Above 1/2	Is considered a "solid" ground fault. Power is reduced to zero.

The other ground leakage detectors will not derate locomotive performance, but when grounds are detected, alarms will be sounded and faults will be logged on the Diagnostic Display Panel.

MOTOR CUT-OUT SWITCHES (Fig. 3)

Traction motors can be cut out manually or automatically. Manual cut out is done with individual Motor Cut-Out switches on the EC panel. Automatic cut out is done by the microcomputer control if a fault condition such as excessive current or too great a rate of change of current (Motor Flashover) is detected.

CAUTION: *It is recommended that motor only be manually cut out when the Engine Control switch is in START or ISOLATE position (unit isolated) and the Throttle handle is in IDLE.*

NOTE: *Speed sensors do not need to be cut out on cut out motors. When the Motor Speed Sensor switch is in the CUT-OUT position, the speed signals from the speed sensors on motors which are cut out are ignored.*

When a motor or motors are cut out, total power available for traction is adjusted as follows:

Motors Cut-Out	Horsepower Available for Input for Traction Per Model			
	Dash 8-32B (B32)	Dash 8-40B (B39)	Dash 8-32C (C32)	Dash 8-40C (C39)
All IN	Full HP	Full HP	Full HP	Full HP
1 Out	See Note 1	See Note 1	Full HP	Full HP
2 Out	See Note 1	See Note 1	Full HP	Full HP
3 Out	See Notes 1 and 2	See Notes 1 and 2	See Note 1	See Note 1
4 Out	See Notes 1 and 2	See Notes 1 and 2	See Note 1	See Note 1
5 Out	N/A	N/A	See Notes 1 and 2	See Notes 1 and 2
6 Out	N/A	N/A	See Notes 1 and 2	See Notes 1 and 2

NOTE 1: Horsepower available for input for traction is limited to 1021 horsepower per each traction motor CUT IN.

NOTE 2: Speed sensor inputs from at least two traction motors are required for locomotive to load.

If any motor is cut out on a locomotive, Dynamic Braking and Self-Load on that locomotive are cut out.

OIL AND WATER TEMPERATURE AND PRESSURE

Horsepower will be derated if one of the following conditions exist:

ALARMS, SAFEGUARDS, POWER DERATIONS AND SHUTDOWNS

Condition	Resulting Load Limit
Oil Temp. below 90 F	Engine RPM is Notch 1 Maximum Power is Notch 1
Oil Temp. below 140 F	Engine RPM is Notch 4 Maximum Power is Notch 4
Oil or Water Temp. between 225 and 239 F	Power is derated from no deration at 225 F to full deration at 235 F
Oil or Water Temp. above 225 F	Engine RPM goes to Notch 8. All power is removed. (Engine returns to requested Notch speed when temp. drops to 218 F.) **NOTE:** *A special combination of sensors together with a built-in program senses operation in a tunnel and permits oil or water temperature to rise to 10 F above these limits for 10 minutes before action is taken.*
Oil or Water Temp. above 240 F	Engine RPM goes to Notch 1.
Low Oil or Water Pressure	Low oil and water pressure are monitored by the engine control governor. If low oil or water pressure is detected: In Motoring or Self-Load - Power is reduced by one third. In Dynamic Brake - Dynamic Brake is nullified if engine speed drops below normal.

OVERSPEED - ENGINE SHUTDOWN (Fig. 32)

In the event the diesel engine overspeeds to 1160 rpm, the engine, radiator fans and the equipment blowers are shut down automatically.

After an overspeed shutdown of the engine, move the EC switch to START.

Reset the overspeed link, pull the layshaft lever toward you until a click is heard. Pushing on the layshaft provides more fuel to the cylinders during cranking, if desired. Note that the location of the lever prevents inadvertent manual overspeeding of the engine beyond the trip setting. Once the device trips, manual control of the fuel racks is eliminated.

Proceed to start the engine as described under Starting Engine section. If it overspeeds again, do not restart the engine.

> **CAUTION:** *During freezing weather, protect the engine cooling system according to railroad instructions.*

OVERSPEED - LOCOMOTIVE

When a locomotive equipped with overspeed protection exceeds the maximum permissible speed, as specified by customer, an Overspeed application is initiated.

1. The overspeed whistle blows.
2. In about five seconds, a Penalty brake application is initiated if train speed has not been reduced sufficiently. See air brake regulation for proper procedure.

PCS SWITCH OPERATION (Fig. 9)

The Pneumatic Control Switch (PCS) is operated from the air brake system. During a safety control Penalty or Emergency brake application, this switch opens. Operation of this switch will affect engine speed, available locomotive power and light the "PCS OPEN" (White) light at the operator's position.

Operating options selected by the railroad will determine how the locomotive control system will react to PCS operation. See railroad operating rules and the locomotive air piping and electrical schematic diagram for application to specific locomotives.

To reset the PCS switch:

1. Move the Throttle handle to IDLE.

NOTE: *If the PCS switch has tripped while in dynamic braking, the Dynamic Braking handle must be returned to OFF to reset the circuit.*

2. Move the Automatic Brake Valve handle to SUPPRESSION.
3. Depress the Safety Control foot pedal (if used). (When the application pipe builds-up to normal pressure, PCS will reclose.)
4. Move the Automatic Brake Valve handle to RELEASE.

POWER LIMIT SWITCH (Fig. 2)

When the Power Limit switch is closed, Engine RPM is limited to Notch 7 and Maximum Power is limited to Notch 7. (Omission of Power Limit Switch is an option.)

SAFETY CONTROL FOOT PEDAL (Fig. 2)

A foot pedal, if installed, is located at the operator's position. The pedal must be depressed at all times during locomotive operation. If the operator's foot is removed from the pedal for more than five seconds, the brakes will apply at the SERVICE rate. For further description, see Safety Controls section.

WHEELSLIP

Axle speed is continuously monitored by the locomotive computers. The axle (or wheel) speed of all axles are compared. If the differential in speeds is greater than a preset limit, wheelslip action will take place as follows:

Wheelslip in Motoring

If differential speeds are greater than the preset limits, a voltage limit is applied to the output of the alternator to keep the differential speed within the specified limit.

If high axle (or wheel) accelerations are measured, indicating a synchronous slip is occurring, the alternator output voltage will also be limited. This voltage limit restriction will be applied proportional to available wheel rail adhesion.

A locked axle or motor overspeed condition will cause output power to go to zero, accompanied by a trainlined wheelslip indication. Power will be applied per the engine load rate schedule when the condition corrects.

ALARMS, SAFEGUARDS, POWER DERATIONS AND SHUTDOWNS

Sand

If poor wheel rail adhesion causes the output of the locomotive to fall below a preset percentage of that requested by the Throttle position, or if a large differential in traction motor speeds exist or if a synchronous slip is detected, sand will be automatically applied.

Wheelslide Braking

In braking, the amount of correction is determined by the amount of slide and is accomplished in several stages:

Stage 1 - Sand (Dynamic Braking)

Automatically apply sand to the leading axles on this locomotive if a small difference in motor speeds is detected. Sanding continues for three seconds after the slip is corrected.

Stage 2 - Small Power Reduction (Dynamic Braking)

When the Stage 1 limit is exceeded, a small power reduction goes into effect and sanding continues.

Stage 3 - Moderate Power Reduction (Dynamic Braking)

When Stage 2 limit of wheelslip is exceeded, a moderate power reduction goes into effect and sanding continues.

Stage 4 - Complete Power Removal (Dynamic Braking)

If a large difference in wheel speeds is detected, a quick power output removal accompanied by a trainlined wheelslip indication results.

The following checks and inspections should be made in accordance with railroad rules:

BEFORE BOARDING LOCOMOTIVE

1. Inspect for broken, worn, loose or dragging parts (brake rigging, brake shoes, wheels, traction motor commutator covers, etc.).
2. Check for leaks from outside piping.
3. Properly position all drain and cut-out cocks.
4. Check the proper connections for air hoses and jumper cables (if in multiple with other units).
5. Check the fuel supply on the fuel tank sight glass.

AFTER BOARDING LOCOMOTIVE

1. Remove rags, tools, etc., from moving parts and electrical equipment WITH ENGINE SHUTDOWN.
2. Check the diesel engine lubricating-oil supply. Oil level should indicate FULL on the dipstick with the engine shut down or at IDLE. The dipstick is located on the side of the engine near the lube-oil fill and is marked LOW and FULL, Fig. 27.
3. Check the governor oil supply. The sight glass on the governor should be full of oil. After engine is started, the oil level must be at the mark on the sight glass, Fig. 25.
4. Check the air compressor lubricating-oil level.

WARNING: *Open Local Control Circuit Breaker (LCCB) to prevent air compressor motor from starting while servicing the air compressor.*

5. Check the cooling water supply. Be sure the water drain valve is closed.
6. Check that the diesel-engine overspeed device is reset, Fig. 31.
7. Check that the engine barring-over device is removed from the engine and cover is mounted in place.
8. Check that the following air cut-out cocks are open:

   a. Air Compressor Governor
   b. Control Air
   c. Safety Control (if used)
   d. Bell, Horn and Window Wiper
   e. Overspeed Control (if used).

9. Check that the brake-pipe angle cocks are properly positioned.
10. The brake valve pilot cut-out cock (double-heading cock) on the 26L air brake system should be properly positioned.
11. The MU2A valve or dual ported cut-out cock must be positioned according to the location of the unit in the locomotive consist.
12. Check the positions of the Automatic and Independent Brake Valve handles. The Automatic Brake Valve handle should be removed on all Trail units, and the Independent handle should be in RELEASE if not removable.
13. Move the Engine Control switch to START.
14. Properly position the MU Headlight Selector switch.
15. Check that the Throttle handle is in IDLE and the Selector handle is in OFF.
16. Check that the dead-engine cock is closed.

STARTING ENGINE

1. Perform operations as in <u>Before Boarding Locomotive</u> and <u>After Boarding Locomotive</u> sections.
2. If the engine has been stopped for a considerable period of time, or if a quantity of rain has entered the stack, the cylinders should be cleared of fuel or water accumulation before starting the engine.

   Proceed as follows:

   a. Apply the engine barring-over device, and back off the compression relief plugs on the left side of each cylinder.
   b. Rotate the engine at least two complete revolutions by use of the engine barring-over device.
   c. Remove the barring-over device from the engine, and tighten all compression relief plugs before cranking.

NOTE: *Cover for barring-over feature must be securely mounted, otherwise engine cannot be cranked. See <u>Barring-Over Switch</u> section.*

3. Check that the emergency stop feature is nullified (Throttle handle in IDLE).
4. Close the Battery switch located behind the door under the EC panel.

PREPARATION FOR OPERATION

5. Turn on all applicable circuit breakers in the top row of breakers on the EC panel.

6. Turn on ALL circuit breakers in the second row of breakers on the EC panel.

NOTE: *When starting engines of several locomotives in a multiple-unit consist, start engines one at a time. Close the Control circuit breaker only on one unit at a time. When all engines are running, close the Control circuit breaker on the Lead unit only, open all others.*

7. Check the Diagnostic Display for any fault messages. If the display says "Won't Crank," the unit will not attempt to crank.

8. Place the Engine Control (EC) switch in the START position.

9. At the Start Station, located near the engine, turn the Start switch to the PRIME position. Hold until solid fuel shows in the sight glass.

10. Turn the switch to the START position and hold until the engine starts.

NOTE: *There will be a 2 to 4 second delay between the time the switch is placed in the START position and the diesel engine starts to rotate.*

NOTE: *If proper engine lube-oil pressure does not build up within approximately 40 seconds, the governor will shut off fuel and prevent the engine from running.*

> **CAUTION:** *Do not discharge the battery excessively by repeated attempts to start. If the first two or three tries are unsuccessful, recheck the starting procedure.*

BEFORE MOVING LOCOMOTIVE

1. Turn the Engine Control switch to RUN.
2. Make an air brake test and other checks in accordance with railroad regulations.
3. Check the main reservoir air pressure according to railroad rules.
4. Check the control air pressure. Normal pressure is 80 psi.
5. Make an Independent air brake application. Release the handbrake and remove any blocking of the wheels.
6. Allow time for the engine cooling water to warm up before moving the locomotive in accordance with railroad rules. Also see ALARMS, SAFEGUARDS, POWER DERATIONS AND SHUTDOWNS section of this manual.
7. Check the Diagnostic Display panel for any fault messages. It should say "Ready."

FASTER AIR PUMPING

To provide faster air pumping on locomotive, when reservoirs have been drained or after the locomotive has been coupled to a train, proceed as follows:

1. Leave the Generator Field circuit breaker in the OFF position.
2. Close the Control breaker on the Engine Control panel.
3. Insert the Reverse handle.
4. Move the Throttle handle to IDLE/Notch 1. At IDLE/Notch 1, the air compressor motor drives the air compressor at twice engine speed.

NOTE: *For optimum air pumping, Throttle handle should be placed in NOTCH 1.*

NOTE: *If the main reservoir air pressure is above 130 psi and is not rising, increasing the engine speed will not raise the pressure.*

COLD WEATHER ENGINE STARTING/WARM-UP

During cold weather conditions, when a locomotive has been shutdown for a period of time, locomotive horsepower will automatically be derated until the lubricating oil temperature reaches a predetermined level. This special warm-up period is required to avoid equipment failure from thermal or overload strain. See ALARMS, SAFEGUARDS, POWER DERATIONS AND SHUTDOWNS section of this manual.

MOVING A TRAIN

1. Close the Generator Field circuit breaker on the control console.
2. Move the Reverse handle to the desired direction of movement.
3. Place foot on the Safety Control foot pedal (if used) and release the brakes completely. Several minutes may be required to release the brakes, depending on the length of the train.
4. Advance the Throttle handle.
5. The Throttle handle has notches (IDLE up to Notch 8), with each successive notch representing an increase in power, or locomotive tractive effort.

Starting a train depends on type, length, weight, grade, condition of rail and amount of slack in the train. This locomotive is designed to have easily controlled tractive effort build-up characteristics, with the tractive effort in each notch limited to definite values as the Throttle handle is moved from the lowest to the highest notch. The operator can easily control the amount of tractive effort required to start and accelerate a particular train. Speed can be controlled as desired by reducing or increasing the Throttle handle position.

STOPPING A TRAIN

Move the Throttle handle to IDLE, and apply the dynamic or air brakes according to railroad regulations. Also see Applying Dynamic Brakes. If leaving the operator's position after the train has stopped, move the Reverse handle to OFF.

OPERATING PROCEDURES

> **CAUTION:** *The control system of this locomotive will delay movement from power to dynamic braking. If however, other locomotives in the consist do not have this feature, to prevent equipment damage when changing from power to dynamic braking or from dynamic braking to power, pause 10 seconds with the Throttle handle at IDLE and Dynamic Brake handle in OFF.*

REVERSING LOCOMOTIVE

1. Bring the locomotive to a full stop.
2. Move the Reverse handle to the opposite direction.
3. Release the brakes.
4. Advance the Throttle handle.

PASSING THROUGH WATER

Do not exceed two or three mph if there is water over the rails. Do not pass through water that is over 2.5 in. above the top of the rail.

PASSING OVER RAILROAD CROSSINGS

Do not pass over railroad crossings at full power, or traction motor flashover may result. Reduce power by moving the Throttle handle to Notch 5, or below, while all units are passing over the crossing.

STOPPING ENGINE

1. Move the Throttle handle to IDLE.

> **CAUTION:** *After a locomotive has operated under full load for a considerable period of time, allow the engine(s) to run at IDLE for at least five minutes before shutting down. Otherwise, immediate shutdown after such operation could be harmful to some engine components requiring brief idling time.*

2. Open the Generator Field circuit breaker on the control stand.
3. Move the Engine Control switch to START.
4. Press the Engine Stop button on the Engine Control panel or at the Engine Start Station.
5. To shut down all engines when in multiple-unit operation, with the Reverser handle in place, move the Throttle handle to the SHUTDOWN position on the Master Controller. The Throttle handle must be in IDLE before attempting to start the engine.

NOTE: *On some older units this will not turn off the fuel pumps. Pushing the STOP button on each unit will turn them off.*

6. Secure the locomotive in accordance with railroad rules and procedures.

BEFORE LEAVING LOCOMOTIVE

1. Apply the handbrake and release the air brakes after uncoupling from the train.

NOTE: *On three-axle floating bolster trucks with low-hung brake cylinders, a "QR," or quick-release valve is provided which removes the air in the one brake cylinder that is in the handbrake system. The handbrake chain must trip the stem of the QR valve and no trapped air is permitted in this brake cylinder; otherwise, if the locomotive air pressure leaks off, the locomotive can roll down the track unattended.*

2. Leave the Throttle handle in IDLE.
3. Close the windows and doors.
4. Open all switches and circuit breakers as described in <u>Control Console Equipment</u> and <u>Engine Control Panel</u> sections of this manual.
5. Open the Battery switch.
6. In freezing weather, precautions must be taken to see that the cooling water does not freeze. See DRAINING COOLING WATER SYSTEM section, and follow railroad rules for this situation.

SAFETY CONTROLS

The safety control (if installed) consists of a foot-pedal operated air valve, whistle and a cut-out cock. Except when the locomotive is stopped and locomotive brakes are applied, the operator must keep the Safety Control foot pedal depressed at all times. This prevents Safety Control brake application.

After a Penalty brake application has occurred, normal locomotive operation is restored in the following manner:

1. Move the Throttle handle to IDLE.
2. Move the Automatic Brake Valve handle to SUPPRESSION.
3. Depress the Safety Control foot pedal.
4. After the application pipe has built up to normal pressure, move the Automatic Brake Valve handle to RELEASE.

NOTE: *Other forms of safety control may be provided. See railroad rules for specific procedures.*

DYNAMIC BRAKE OPERATION

Dynamic braking is applied to the locomotive only.

APPLYING DYNAMIC BRAKES

Applying dynamic braking is done in the following manner:

NOTE: *Dynamic brake cannot be applied on a locomotive which has any traction motor manually or automatically cut out.*

1. Move Throttle handle to IDLE.
2. Move the Dynamic Brake handle to SET-UP position; pause, then advance the handle into the BRAKING sector as desired.
3. After the slack is bunched, manipulate the Dynamic Braking handle until the desired braking effort is obtained. Observe and correct braking effort during the initial period of Dynamic Brake application.

> **CAUTION:** *Prolonged operation of dynamic braking in Notch 8 at speeds above 61 miles per hour can cause increased maintenance requirements of traction motors.*

The amount of braking effort obtainable varies with the position of the Dynamic Braking handle for various speeds. Maximum braking effort is obtained in the FULL BRAKING position at speeds of 22 to 30 mph, depending on locomotive gearing.

When a locomotive is equipped with extended range dynamic braking, a series of peak braking efforts will occur down to about 8 mph. If independent air brakes are applied when dynamic braking is in effect, only minimum dynamic brake will be obtained.

NOTE: *Wheelslip warning may occur while in dynamic braking. This indicates wheels are sliding. Sand is applied automatically to the wheels of the sliding unit. Reduce the Braking handle position until the warning stops.*

USE OF AIR BRAKES DURING DYNAMIC BRAKING

NOTE: *If independent air brake pressure is applied above 20 psi during dynamic braking, the dynamic braking effort will immediately go to, and remain at, MINIMUM braking until independent is released.*

When necessary, the automatic air brake may be used in conjunction with the dynamic brake. Automatic air brakes will apply on the train but not on the locomotive. If the Automatic Air Brake handle is moved to the EMERGENCY position, the dynamic brake is removed and brakes on the locomotive, as well as those on the train, go into Emergency application.

The Dynamic Brake Magnet Valve (DBM) nullifies an Automatic air brake application on the locomotives when dynamic braking is being used. This same interlock will release an Automatic application on the locomotives when dynamic brakes are set-up, and prevents reapplication of the automatic brake on the locomotive after release of the dynamic brake.

NOTE: *Optional arrangements, if selected by the railroad, may allow automatic brakes on the locomotive to reapply after the release of dynamic brake or may eliminate the Dynamic Brake Magnet Valve (DBM). If DBM is eliminated, automatic brakes will not be released when dynamic brake is applied. Operate air brake and dynamic brake in accordance with railroad operating procedures.*

RELEASE OF DYNAMIC BRAKING

Release dynamic braking by moving the Dynamic Braking handle to the OFF position.

MULTIPLE-UNIT OPERATION

OPERATING AS A LEADING UNIT

To operate the locomotive as a Lead unit of a consist, first make the necessary preliminary preparations for operation then proceed as follows:

Air Equipment Set-Up

1. Insert the Automatic Brake Valve handle in the HANDLE OFF position.
2. Depress the handle of the brake-valve pilot cut-out cock and move it to the IN position.
3. Depress the handle of the MU2A valve and move it to the LEAD/DEAD position or move the handle of the dual ported cut-out cock to the IN/OPEN position.
4. Move the Independent Brake Valve handle to the FULL APPLICATION position.
5. Test the air brake in accordance with railroad rules.

Operating Unit - Electrical Set-Up

1. Close the Generator Field circuit breaker on the control stand. (The Control circuit breaker must be closed on the Lead unit only.)
2. Close the Dynamic Brake circuit breaker (if so equipped).
3. Close the Control circuit breaker.
4. Close all circuit breakers on the Engine Control (EC) panel.
5. Move the MU Headlight Set-Up switch to the required position.

6. Insert the Reverse handle into the Controller.

7. Move the Reverse handle to the desired direction.

8. Operate the locomotive in accordance with operating procedure.

OPERATING AS A TRAILING UNIT

Air Equipment Set-Up

1. Make a Full Service application with the Automatic Brake Valve handle.

2. Move the brake valve pilot cut-out (double-heading) cock to the OUT position.

3. Move the Automatic Brake Valve handle to the HANDLE OFF position and remove the handle.

4. Place the Independent handle in RELEASE position.

5. Move the MU2A valve to LEAD/DEAD position, or if the dual ported cut-out cock is used, move the handle to the OUT/CLOSED position.

Electrical Set-Up

1. Move the Reverse handle to OFF and remove the handle.

2. Open the Generator Field, Control, Engine Run and Dynamic Brake circuit breakers on the control stand.

3. The top row of circuit breakers on the Engine Control (EC) panel can be turned OFF for Trail operation. Second row of breakers MUST BE ON for Trail operation. The Running Lights circuit breaker may be positioned as desired.

4. Place the MU Headlight Set-Up switch in the proper position.

CHANGING OPERATING ENDS

To change operating control from the cab of one locomotive unit to the cab of another, proceed as follows:

Vacating Unit - Air Equipment Set-Up

1. Make a Full Service brake-pipe reduction.

2. Allow time for all air blowing sounds to stop; then depress the handle of the brake valve pilot cut-out cock and move it to the OUT position.

3. Place the Automatic Brake Valve handle in the HANDLE OFF position and remove; place the Independent Brake Valve handle in the RELEASE position.

4. Depress the handle on the MU2A valve, and move it to TRAIL position, or the dual ported cut-out cock to the OUT/CLOSED position.

Vacating Unit - Electrical Set-Up

1. Move the Reverse handle to OFF, and remove the handle.

2. Open the Generator Field, Control, Engine Run and Dynamic Brake circuit breakers on the control stand.

3. The top row of circuit breakers on the Engine Control (EC) panel can be turned OFF for Trail operation. Second row of breakers MUST BE ON for Trail operation. The Running Lights circuit breaker may be positioned as desired.

4. Move the MU Headlight Set-Up switch to the required position.

Operating Unit - Air and Electrical Equipment Set-Up

Set-up the air brakes and electrical equipment on the operating unit as described in Operating As a Leading Unit "Air Equipment Set-Up" and "Electrical Equipment Set-Up" sections.

TO OPERATE WITH OTHER TYPES OF UNITS

This locomotive is equipped with a traction motor thermal simulator which computes traction motor temperatures. This simulator will reduce locomotive output as required to protect the traction motors.

If the units in the locomotive consist are geared for differing maximum speeds, do not run at speeds in excess of that recommended for the unit having the lowest maximum permissible speed.

Similarly, do not operate at low speeds long enough to exceed the specified traction motor ratings on any of the units in the locomotive consist. A locomotive with high horsepower per axle will develop more tractive effort at any given speed than will units of lower horsepower per axle and will, therefore, tend to overload sooner at lower speeds.

When the leading unit is slipping excessively, the Power-Limit switch (if so equipped) can be moved to NOTCH 7 to reduce the power on this unit while the Trailing units are operating at full power. This will reduce the tractive effort on the Leading unit and will usually improve the ability of the locomotive to hold the rail under bad rail conditions.

BRAKE PIPE LEAKAGE TEST

A brake-pipe leakage test can be performed in the following manner:

With the brake system fully charged and with the brake-valve pilot cut-out cock in the IN position, move the Automatic Brake Valve handle promptly toward the SERVICE position until the equalizing reservoir pressure has been reduced 15 psi; then stop and leave the handle in this position.

As soon as the brake-pipe pressure has reduced to the level of the equalizing reservoir pressure (continuous blow from brake-valve exhaust), depress the Brake-Valve Pilot Cut-Out Cock handle and move it to the OUT position. Immediately observe the brake-pipe gage, and time the pressure drop in accordance with railroad rules.

At the completion of the brake-pipe leakage test, move the Brake Valve handle further toward the SERVICE position, and reduce the equalizing reservoir pressure slightly below the brake-pipe pressure. The brake may later be released by returning the Brake Valve handle to the RELEASE position.

DEAD HEADING (DEAD-IN-TRAIN)

1. Place the Independent Brake Valve handle in the RELEASE position and the Automatic Brake Valve handle in the HANDLE OFF position.

2. Depress the Brake Valve Pilot Cut-Out handle and move to the OUT position.

3. Depress the handle of the MU2A valve and move to the LEAD/DEAD position. On units equipped with a dual-ported cut-out cock, place the cock in the IN/OPEN position.

CAUTION: *To avoid wheel flats, drain main reservoirs of unit 40 psi below the brake pipe pressure used on the train to which the locomotive will be coupled.*

4. Open the dead-engine cock.

CLYDE W. SCOTT
P.O. BOX 507
BOTHELL, WA 98041

CLYDE W. SCOTT
P.O. BOX 507
BOTHELL, WA 98041

CLYDE W. SCOTT
P.O. BOX 507
BOTHELL, WA 98041